地理信息系统应用与开发丛书

# 地理信息系统开发与编程实验教程

李进强　编著

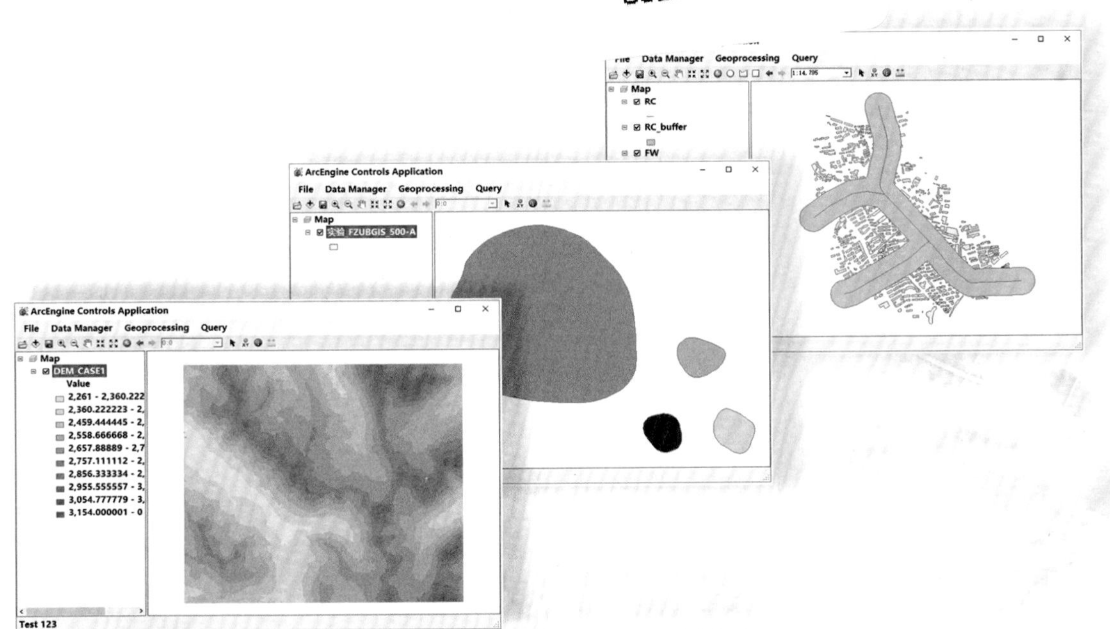

WUHAN UNIVERSITY PRESS
武汉大学出版社

**图书在版编目(CIP)数据**

地理信息系统开发与编程实验教程/李进强编著. —武汉:武汉大学出版社,2018.11
地理信息系统应用与开发丛书
ISBN 978-7-307-16244-0

Ⅰ.地… Ⅱ.李… Ⅲ. 地理信息系统—教材 Ⅳ.P208

中国版本图书馆 CIP 数据核字(2018)第 239193 号

责任编辑:鲍 玲　　责任校对:李孟潇　　装帧设计:韩闻锦

---

出版发行:**武汉大学出版社** (430072 武昌 珞珈山)
(电子邮件:cbs22@whu.edu.cn 网址:www.wdp.com.cn)
印刷:湖北金海印务有限公司
开本:787×1092 1/16　印张:14.75　字数:347 千字　插页:1
版次:2018 年 11 月第 1 版　2018 年 11 月第 1 次印刷
ISBN 978-7-307-16244-0　定价:36.00 元

---

# 前　言

本书以 ArcGIS Engine10. 5+Visual Studio 2015 组件式开发为主线，通过 21 个典型 GIS 开发实验案例，介绍了地理信息系统开发框架设计，空间信息可视化，空间查询统计与分析，空间数据管理等方面初步开发技术与方法。

各部分按照以下顺序展开：

目的与要求：阐述通过本实验达到的效果；

实验原理：对每部分所涉及的 ArcGIS Engine 接口、实现类，以及对应的属性和方法进行了详细讲解；

内容与步骤：主要介绍实验步骤和参考代码，对技术难点内容作出恰当解释。

编译与测试：配套实验数据可供读者测试所编程序的正确性。

思考与练习：提出了若干同类思考问题和扩展练习题，供进一步提升学习使用。

为简化学习难度，作者在编写时尽量简化界面元素，并尽量减少多技术关联（如读者需要了解实用化的开发技术，请参考作者编著的《基于 ArcGIS Engine 地理信息系统开发技术与实践》一书），使得每个实验相互独立又有联系，帮助读者掌握面向对象程序设计的基本知识。

本书可作为地理科学、资源环境科学等相关专业及开展“地理信息系统设计与编程”课程教学的实验教材，也可供其他相关人员自学参考。

由于作者水平有限，再加上编写时间仓促，书中错漏之处在所难免，敬请读者批评指正。作者邮箱：1361639771@ qq. com。

# 目　　录

# 实验一　投影计算程序设计(C#增强)

## 一、目的与要求

通过投影正反算的编程，加深对面向对象程序设计的理解，巩固复习有关 C 语言的基础知识，同时提高学生运用计算机技术解决实际问题的能力。

本次实验的主要任务是：每人独立完成 Gauss、Lambert 投影正反算算法设计、计算编程、程序调试等，并通过实例验证程序的正确性。

实验过程中，同学们应当复习《地图投影学》或《控制测量学》有关内容。

## 二、实验原理

### 1. Gauss 投影计算公式

**(1)基本公式**

Gauss 投影正算公式(精度精确至 0.001m，角度 $B$，$L$ 以弧度为单位)：

$$\left.\begin{aligned} x &= X + \frac{N}{2}\sin B\cos B L^2 + \frac{N}{24}\sin B\cos^3 B(5 - t^2 + 9\eta^2 + 4\eta^4)L^4 + \\ &\quad \frac{N}{720}\sin B\cos^5 B(61 - 58t^2 + t^4)L^6 \\ y &= N\cos B L + \frac{N}{6}\cos^3 B(1 - t^2 + \eta^2)L^3 + \\ &\quad \frac{N}{120}\cos^5 B(5 - 18t^2 + t^4 + 14\eta^2 - 58\eta^2 t^2)L^5 \end{aligned}\right\}$$

Gauss 投影反算公式：

$$\left.\begin{aligned} B &= B_f - \frac{t_f}{2M_f N_f}y^2 + \frac{t_f}{24M_f N_f^3}(5 + 3t_f^3 + \eta_f^2 - 9\eta_f^2 t_f^2)y^4 - \\ &\quad \frac{t_f}{720M_f N_f^5}(61 + 90t_f^2 + 45t_f^4)y^6 \\ L &= \frac{1}{N_f\cos B_f}y - \frac{1}{6N_f^3\cos B_f}(1 + 2t_f^2 + \eta_f^2)y^3 + \\ &\quad \frac{1}{120N_f^5\cos B_f}(5 + 28t_f^2 + 24t_f^4 + 6\eta_f^2 + 8\eta_f^2 t_f^2)y^5 \end{aligned}\right\}$$

子午线收敛角计算公式：

$$\gamma = \sin B \cdot L + \frac{1}{3}\sin B\cos^2 B \cdot L^3(1 + 3\eta^2 + 2\eta^4) + \frac{1}{15}\sin B\cos^4 B \cdot L^5(2 - t^2) + \cdots$$

**(2) 子午线长度计算公式**

$$X = a_0 B - \frac{a_2}{2}\sin 2B + \frac{a_4}{4}\sin 4B - \frac{a_6}{6}\sin 6B + \frac{a_8}{8}\sin 8B$$

$$\left.\begin{aligned} a_0 &= m_0 + \frac{m_2}{2} + \frac{3}{8}m_4 + \frac{5}{16}m_6 + \frac{35}{128}m_8 + \cdots \\ a_2 &= \frac{m_2}{2} + \frac{m_4}{2} + \frac{15}{32}m_6 + \frac{7}{16}m_8 \\ a_4 &= \frac{m_4}{8} + \frac{3}{16}m_6 + \frac{7}{32}m_8 \\ a_6 &= \frac{m_6}{32} + \frac{m_8}{16} \\ a_8 &= \frac{m_8}{128} \end{aligned}\right\}$$

**(3) 子午圈卯酉圈曲率半径计算公式**

$$N = \frac{a}{W}, \quad M = \frac{c}{V^3}$$

$$\left.\begin{aligned} W &= \sqrt{1 - e^2\sin^2 B} \\ V &= \sqrt{1 + e'^2\cos^2 B} \end{aligned}\right\}$$

**(4) 椭球参数计算公式**

$$c = \frac{a^2}{b}, \quad t = \tan B, \quad \eta^2 = e'^2\cos^2 B$$

$$e = \frac{\sqrt{a^2 - b^2}}{a}, \quad e' = \frac{\sqrt{a^2 - b^2}}{b}$$

$$m_0 = a(1 - e^2), \quad m_2 = \frac{3}{2}e^2 m_0, \quad m_4 = \frac{5}{4}e^2 m_2,$$

$$m_6 = \frac{7}{6}e^2 m_4, \quad m_8 = \frac{9}{8}e^2 m_6$$

2. Lambert 投影计算公式

**(1) 正算公式**

$$\begin{cases} X = r_0 - r\cos(\theta) \\ Y = r\sin(\theta) \end{cases}$$

说明：$r$ 为纬线投影半径（$r_0$ 为原点纬线投影半径），$\theta$ 为计算点的子午线与中央子午线投影后的夹角。

$$r = aFt^n$$

$$r_0 = aFt_0^n$$

$$\theta = n(L - L_0)$$

其中，$n$，$F$ 值由下式计算：

$$n = \frac{\ln(m_{B_1}/m_{B_2})}{\ln(t_{B_1}/t_{B_2})}$$

$$F = m_{B_1}/(nt_{B_1}^n)$$

其中，$m_{B_1}$，$m_{B_2}$分别为标准维度 $B_1$，$B_2$处的 $m$ 值。

其中 $t_{B_1}$，$t_{B_2}$，$t_0$分别为标准维度 $B_1$，$B_2$和原点维度 $B_0$处的 $t$ 值。

计算公式如下：

$$m = \frac{\cos(B)}{\sqrt{1 - e^2 \sin^2(B)}}$$

$$t = \tan\left(\frac{\pi}{4} - \frac{B}{2}\right) / \left(\frac{1 - e\sin(B)}{1 + e\sin(B)}\right)^{\frac{e}{2}}$$

**(2)反算公式**

$$B = \frac{\pi}{2} - 2\arctan\left[t'\left(\frac{1 - e\sin(B)}{1 + e\sin(B)}\right)^{\frac{e}{2}}\right]$$

$$L = \frac{\theta'}{n} + L_0$$

$$t' = \left(\frac{r'}{aF}\right)^{\frac{1}{n}}$$

$$\theta' = \arctan\frac{Y}{r_0 - X}$$

$$r' = \mathrm{Sign}(n) \cdot \sqrt{Y^2 - (r_0 - X)^2}$$

说明：

公式中 $n$，$r_0$与坐标正算方法相同。

$r'$，$\theta'$分别表示由直角坐标计算得出的纬线投影半径和经线与中央子午线的投影角。

Sign($n$)为 $n$ 的符号函数，结果为+1 或-1。

## 三、实验环境与数据

①编程环境：VS 2015 C#。

②实验数据：

Lambert 投影要求实现以下几组坐标之间的互相转换，详见表 1-1。

表 1-1 **利用 Lambert 投影实现的几组坐标之间的转换**

| 椭球 | $B$ | $L$ | $L_0$ | $X$ | $Y$ |
|---|---|---|---|---|---|
| 克拉索夫斯基 | 43°18′17. 6562 | 87°57′11. 0581 | 105° | 5294592. 403 | -1321241. 192 |
| 克拉索夫斯基 | 35°34′11. 5623 | 106°23′17. 1521 | 105° | 4319291. 848 | 123137. 909 |

其中：原点纬度：0°；第一标准纬度：30°；第二标准纬度：62°。

Gauss 投影要求实现以下几组坐标之间的互相转换，见表 1-2。

表 1-2 **利用 Gauss 投影实现的几组坐标之间的转换**

| 椭球 | $B$ | $L$ | $L_0$ | $X$ | $Y$ |
|---|---|---|---|---|---|
| 克拉索夫斯基 | 51°38′43. 9023″ | 126°02′13. 1360″ | 123° | 5728374. 726 | 210198. 193 |
| 克拉索夫斯基 | 29°34′16. 5112″ | 106°25′14. 8663″ | 105° | 3273488. 972 | 137682. 377 |
| 1975 年国际椭球 | 51°38′43. 9023″ | 126°02′13. 1360″ | 123° | 5728276. 609 | 210194. 803 |
| 1975 年国际椭球 | 29°34′16. 5112″ | 106°25′14. 8663″ | 105° | 3273431. 384 | 137680. 138 |

1975 国际椭球体参数：

$a$=6378140. 0；

$b$=6356755. 2881575287。

克拉索夫斯基椭球体参数：

$a$=6378245. 0；

$b$=6356863. 0187730473。

## 四、实验内容与步骤

### 1. 接口/基类/辅助结构定义

**(1) GeorefEllipsoid 基类定义**

定义参考椭球体基类 GeorefEllipsoid，提供获取参考椭球体长半轴、短半轴、偏心率等基本计算功能，设计代码如下：

```
public class GeorefEllipsoid
{
    /// <summary>
    /// 椭球体长半轴(单位:米)
    /// </summary>
    private double _a;
    /// <summary>
```

```
///椭球体短半轴(单位:米)
///</summary>
private double _b;
///<summary>
///构造函数
///</summary>
///<param name="A"></param>
///<param name="B"></param>
public GeorefEllipsoid(double A, double B)
{
    _a = A;
    _b = B;
}
///<summary>
///椭球体长半轴
///</summary>
public double a
{
    get { return _a; }
}
///<summary>
///椭球体短半轴
///</summary>
public double b
{
    get { return _b; }
}
///<summary>
///偏心率(a-b)/a
///</summary>
public double f
{
    get { return (_a - _b) /_a; }
}
///<summary>
///第一偏心率
///</summary>
public double e1
{
```

```
            get
            {
                    return Math.Sqrt(a * a - b * b) /a;
            }
        }
        ///<summary>
        ///第二偏心率
        ///</summary>
        public double e2
        {
            get
            {
                return Math.Sqrt(a * a - b * b) /b;
            }
        }
    }
```

**(2) IProjection 接口定义**

定义两个投影计算接口函数：地理坐标投影到直角坐标，直角坐标反算到地理坐标：

```
interface IProjection
{
    Vector2d BLtoVector2d( double B, double L);
    Geographics XYtoGeographics(double X, double Y);
}
```

**(3) Vector2d、Geographics 辅助结构**

Vector2d、Geographics 结构分别表示直角坐标和地理坐标：

```
public struct Vector2d
{
    public double X;
    public double Y;
    public Vector2d(double x, double y)
    {
        X = x;
        Y = y;
    }
}

public struct Geographics
{
    public double B;
```

```
    public double L;

    public Geographics(double lat, double lng)
    {
        B = lat;
        L = lng;
    }
}
```

**(4)Function 静态类实现**

定义静态类 Function，提供“拟十进制”角度和“十进制”角度的转换：

```
public static class Function
{
    public static double DegToDms(double deg)
    {
        double d, m, s;
        s = (deg > 0) ? deg : -deg;
        d = Math.Floor(s);
        s = (s - d) * 60;
        m = Math.Floor(s);
        s = (s - m) * 60;

        s = d + m /100 + s /10000;
        return (deg > 0) ? s : -s;
    }
    //dms to deg
    public static double DmsToDeg(double dms)
    {
        double d, m, s;
        s = (dms > 0) ? dms : -dms;
        d = Math.Floor(s);
        s = (s - d) * 100;
        m = Math.Floor(s);
        s = (s - m) * 100;

        s = (d + (m /60) + s /3600);
        return (dms > 0) ? s : -s;
    }
}
```

### 2. 创建 Gauss 投影计算窗体

使用 VS 2015 模板创建 Windows 应用程序，项目命名为“Projection”，创建窗体 GaussProjection。

①在 GaussProjection 窗体设计界面上，添加 5 个 TextBox、2 个 Button 控件，详见表 1-3。

表 1-3　**GaussProjection 窗体控件命名及 Name 属性**

| 控件 | Name 属性 | 含义 |
| --- | --- | --- |
| TextBox | txtLongitude | 经度 |
| TextBox | txtLatitude | 纬度 |
| TextBox | txtX | Gauss 坐标 X |
| TextBox | txtY | Gauss 坐标 Y |
| TextBox | txtCentreLng | 中央子午线 |
| Button | btnComputation | 正算 |
| Button | btnInverseComputation | 反算 |

效果如图 1-1 所示。

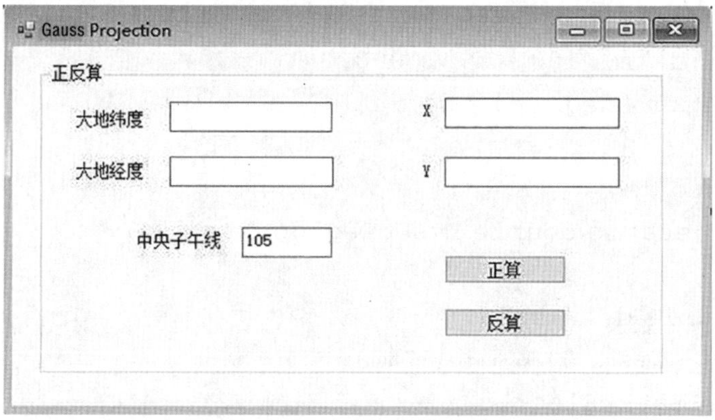

图 1-1　GaussProjection 用户界面

②添加两个常量成员(椭球的长轴，短轴)：

const double _a = 6378140.0;

const double _b = 6356755.2881575287。

③为两个 Button 控件的 Click 事件添加响应函数。

代码如下：

```
public partial classGaussFrm : Form
{
```

```
//1975 国际椭球体参数
const double _a = 6378140.0;
const double _b = 6356755.2881575287;

public GaussFrm()
{
    InitializeComponent();
}

//正算响应函数
private void btnComputation_Click(object sender, EventArgs e)
{
    //从界面获取经度、纬度、中央子午线,同时转换为十进制角度
    double degL = Function.DmsToDeg(double.Parse(txtLongitude.
Text));
    double degB = Function.DmsToDeg(double.Parse(txtLatitude.
Text));
    int iCentreL = int.Parse(txtCentreLng.Text);

    //计算经差,并转为弧度
    double dL = (degL - iCentreL) * Math.PI /180;
    double dB = degB * Math.PI /180;

    //初始化 GaussProjection 对象,调用 BLtoVector2d 函数完成投影正算
    IProjection gp = new GaussProjection(_a, _b);
    Vector2d vet = gp.BLtoVector2d(dB, dL);

    //Y 附加常数 500 公里,取整至 0.001
    double X = Math.Round(vet.X, 3);
    double Y = Math.Round(vet.Y + 500000.0, 3);

    //更新 X、Y 显示控件
    txtX.Text = X.ToString();
    txtY.Text = Y.ToString();
}

//反算响应函数
private void btnInverseComputation_Click(object sender,
```

```
EventArgs e)
        {

            //从界面获取X、Y、中央子午线
            double dX = double.Parse(txtX.Text);
            double dY = double.Parse(txtY.Text) - 500000;
            int  iCentreL = int.Parse(txtCentreLng.Text);

            //初始化GaussProjection对象,调用XYtoGeographics函数完成投影反算
            IProjection gp = new GaussProjection(_a, _b);
            Geographics pos = gp.XYtoGeographics(dX, dY);

            //弧度制经纬度转换为拟十进制,由经差计算经度
             double degL = Function.DegToDms(pos.L * 180 /Math.PI +
iCentreL);
            double degB = Function.DegToDms(pos.B * 180 /Math.PI);

            //取整至0.00000001
            degB = Math.Round(degB, 8);
            degL = Math.Round(degL, 8);

            //更新L,B显示控件
            txtLongitude.Text = degL.ToString();
          txtLatitude.Text = degB.ToString();
        }
    }
```

3. GaussProjection 功能类实现

窗体类用到的功能类是 GaussProjection，该类继承 GeorefEllipsoid，IProjection 实现，代码如下：

```
    public class GaussProjection: GeorefEllipsoid, IProjection
    {
        public GaussProjection(double a, double b)
                :base(a, b)
        {
        }

        ///正算
```

```
public Vector2d BLtoVector2d( double dB, double dL )
{
        //计算辅助函数
    double cosb, sinb, W, V, N, t, g, p;
    cosb = Math.Cos(dB);
    sinb = Math.Sin(dB);
    W = Math.Sqrt(1 - e1 * e1 * sinb * sinb);
    V = Math.Sqrt(1 + e2 * e2 * cosb * cosb);
    N = a /W;
    t = Math.Tan(dB);
    g = e2 * e2 * cosb * cosb;
    p = 1;

    //计算子午线弧长
    double[] A = getMeridianDisParameter();
    double dX = A[0] * dB - (A[2] /2) * Math.Sin(2 * dB) + (A[4]
/4) * Math.Sin(4 * dB) - (A[6] /6) * Math.Sin(6 * dB) + (A[8] /8) *
Math.Sin(8 * dB);

    //计算 X , Y , R
    double l1_ = dL /p;
    double l2_ = l1_ * l1_;
    double l3_ = l2_ * l1_;
    double l4_ = l2_ * l2_;
    double l5_ = l4_ * l1_;
    double l6_ = l4_ * l2_;
    double cosb3w = cosb * cosb * cosb;
    double cosb5w = cosb3w * cosb * cosb;
    double t2 = t * t;
    double t4 = t2 * t2;
    double g2 = g * g;
    double g4 = g2 * g2;

    dX = dX + (N /2) * sinb * cosb * l2_ + (N /24) * sinb *
cosb3w * (5 - t2 + 9 * g2 + 4 * g4) * l4_ + (N /720) * sinb * cosb5w *
(61 - 58 * t2 + t4) * l6_;

    double dY = N * cosb * l1_ + (N /6) * cosb3w * (1 - t2 + g2)
```

```
* l3_ + (N /120) * cosb5w * (5 - 18 * t2 + t4 + 14 * g2 - 58 * g2 * t2)
* l5_;

            return new Vector2d(dX, dY);
        }

        ///反算
        public Geographics XYtoGeographics(double dX, double dY)
        {
         //计算底点纬度
            double bf = get_Bf(dX);

            //计算辅助函数
            double tf = Math.Tan(bf);
            double cosbf = Math.Cos(bf);
            double sinbf = Math.Sin(bf);

            double Wf = Math.Sqrt(1 - e1 * e1 * sinbf * sinbf);
            double Vf = Math.Sqrt(1 + e2 * e2 * cosbf * cosbf);
            double Nf = a /Wf;
            double Mf = Nf /(Vf * Vf);
            double gf = e2 * e2 * cosbf * cosbf;

            //计算 L ,B
            double y2 = dY * dY;
            double y3 = dY * y2;
            double y4 = y2 * y2;
            double y5 = dY * y4;

            double tf2 = tf * tf;
            double tf3 = tf * tf2;
            double tf4 = tf2 * tf2;
            double Nf3 = Nf * Nf * Nf;
            double Nf5 = Nf * Nf * Nf3;
            double gf2 = gf * gf;

            double dB = bf - tf * y2 /(2 * Mf * Nf) + tf * (5 + 3 * tf2 +
gf2 - 9 * gf2 * tf2) * y4 /(24 * Mf * Nf3);
```

```
            double dL = dY /(Nf * cosbf) - y3 * (1 + 2 * tf2 + gf2) /(6 *
Nf3 * cosbf) +
    y5 * (5 + 28 * tf2 + 24 * tf4 + 6 * gf2 + 8 * gf2 *tf2) /(120 * Nf5
 * cosbf);

            return new Geographics(dB, dL);
        }

        ///<summary>
        ///迭代法计算底点纬度
        ///</summary>
        ///<param name="a">长半轴(米)</param>
        ///<param name="b">短半轴(米)</param>
        ///<param name="X">gaussX 坐标</param>
        ///<returns></returns>
        private double get_Bf(double dX)
        {

            double[] A = getMeridianDisParameter();
            double sin2b, sin4b, sin6b, sin8b;

            double be = 0.0;
            double bf = dX /A[0];
            while (Math.Abs(bf - be) > 5e-13)
            {
                be = bf;
                sin2b = Math.Sin(2 * be);
                sin4b = Math.Sin(4 * be);
                sin6b = Math.Sin(6 * be);
                sin8b = Math.Sin(8 * be);
                bf = (dX + A[2] * sin2b /2 - A[4] * sin4b /4 + A[6] *
sin6b /6 - A[8] * sin8b /8) /A[0];
            }

            return bf;
        }
```

```
        /// <summary>
        ///椭球体 A 参数
        /// </summary>
        /// <param name = "a">长半轴(米)</param>
        /// <param name = "b">短半轴(米)</param>
        /// <returns></returns>
        private double[] getMeridianDisParameter()
        {
          //计算 m
            double m0, m2, m4, m6, m8;
            m0 = a * (1 - e1 * e1);
            m2 = (3.0 /2) * e1 * e1 * m0;
            m4 = (5.0 /4) * e1 * e1 * m2;
            m6 = (7.0 /6) * e1 * e1 * m4;
            m8 = (9.0 /8) * e1 * e1 * m6;

            double[] A = new double[9];
            A[0] = m0 + m2 /2 + m4 * 3 /8 +  m6 * 5 /16 + m8 * 35 /128;
            A[2] = m2 /2 + m4 /2 + m6 * 15 /32 + m8 * 7 /16;
            A[4] = m4 /8 + m6 * 3 /16 + m8 * 7 /32;
            A[6] = m6 /32 + m8 /16;
            A[8] = m8 /128;

            return A;
        }
    }
```

### 4. 创建 Lambert 投影计算窗体

在 Projection 项目中创建窗体 LambertProjection。

步骤如下：

①在 LambertProjection 窗体设计界面上，添加 8 个 TextBox、2 个 Button 控件，控件命名及含义详见表 1-4。

表 1-4 **LambertProjection 窗件控件命名及 Name 属性**

| 控件 | Name 属性 | 说明 |
| --- | --- | --- |
| TextBox | txtLongitude | 经度 |
| TextBox | txtLatitude | 纬度 |

续表

| 控件 | Name 属性 | 说明 |
| --- | --- | --- |
| TextBox | txtX. Text | Lambert 坐标 X |
| TextBox | txtY. Text | Lambert 坐标 Y |
| TextBox | txtCentreLng | 中央子午线 |
| TextBox | txtOrgLat | 原点纬度 |
| TextBox | txtStdLat1 | 第一标准纬度 |
| TextBox | txtStdLat2 | 第二标准纬度 |
| Button | btnComputation | 正算 |
| Button | btnInverseComputation | 反算 |

效果如图 1-2 所示。

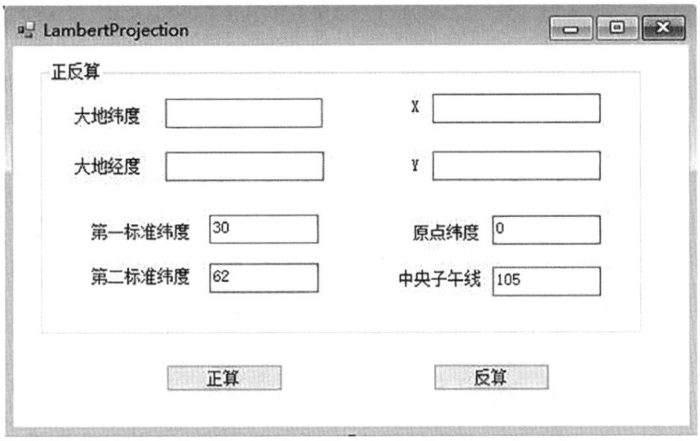

图 1-2　LambertProjection 窗体用户界面

②添加两个常量成员(椭球的长轴，短轴)：

const double _a = 6378245.0;

const double _b = 6356863.0187730473。

③为两个 Button 控件的 Click 事件添加响应函数。

代码如下：

```
public partial class LambertFrm : Form
{
    //克拉索夫斯基椭球体参数
    const double _a = 6378245.0;
    const double _b = 6356863.0187730473;
    public LambertFrm()
```

```
        {
            InitializeComponent();
        }

        private void btnComputation_Click(object sender, EventArgs e)
        {
            //从界面获取经度、纬度、中央子午线等,同时转换为弧度角度
            double degL = Function.DmsToDeg(double.Parse(txtLongitude.
Text));
            double degB = Function.DmsToDeg(double.Parse(txtLatitude.
Text));
            double dL = degL * Math.PI /180;
            double dB = degB * Math.PI /180;

            double iOrgLat = (Math.PI /180) * int.Parse(txtOrgLat.
Text);
            double iCentreLng = (Math.PI/180) * int.Parse(txtCentreLng.
Text);
            double iStdLat1 = (Math.PI /180) * int.Parse(txtStdLat1.
Text) ;
            double iStdLat2 = (Math.PI /180) * int.Parse(txtStdLat2.
Text);

            //初始化 LambertProjection 类,调用 BLtoVector2d 函数完成投影正算
            Geographics org = new Geographics(iOrgLat, iCentreLng);
            IProjection lp = new LambertProjection(_a, _b, org, iStdLat1,
iStdLat2);
            Vector2d vet = lp.BLtoVector2d(dB, dL);

            //X、Y 取整至 0.001m
            double X = Math.Round(vet.X, 3);
            double Y = Math.Round(vet.Y, 3);

            //更新 X、Y 显示控件;
            txtX.Text = X.ToString();
            txtY.Text = Y.ToString();
        }
```

```
        private void btnInverseComputation_Click(object sender, EventArgs e)
        {
            //从界面获取X、Y、中央子午线等,同时将角度转换为弧度角度
            double dX = double.Parse(txtX.Text);
            double dY = double.Parse(txtY.Text);

            double iOrgLat = (Math.PI /180) * int.Parse(txtOrgLat.Text);
            double iCentreLng=(Math.PI/180) * int.Parse(txtCentreLng.Text);
            double iStdLat1 = (Math.PI /180) * int.Parse(txtStdLat1.Text);
            double iStdLat2 = (Math.PI /180) * int.Parse(txtStdLat2.Text);

            //初始化LambertProjection对象,调用XYtoGeographics函数完成投影反算
            Geographics org = new Geographics(iOrgLat, iCentreLng);
            IProjection lp = new LambertProjection(_a, _b, org, iStdLat1, iStdLat2);
            Geographics pos = lp.XYtoGeographics(dX, dY);

            //弧度制经纬度转换为拟十进制
            double degL = Function.DegToDms(pos.L * 180 /Math.PI);
            double degB = Function.DegToDms(pos.B * 180 /Math.PI);

            //取整至0.00000001
            degB = Math.Round(degB, 8);
            degL = Math.Round(degL, 8);

            //更新L,B显示控件;
            txtLongitude.Text = degL.ToString();
            txtLatitude.Text = degB.ToString();
        }
    }
```

5. LambertProjection 功能类实现

窗体类用到的功能类是 LambertProjection，和 GaussProjection 一样，也是通过继承

GeorefEllipsoid，IProjection 实现，代码如下：

```
public class LambertProjection : GeorefEllipsoid, IProjection
{
    ///<summary>
    ///原点纬度
    ///</summary>
    private Geographics _Org;
    ///<summary>
    ///第一标准纬度
    ///</summary>
    private double _stdLat1;
    ///<summary>
    ///第二标准纬度
    ///</summary>
    private double _stdLat2;

    ///<summary>
    ///构造函数
    ///</summary>
    ///<param name="a"></param>
    ///<param name="b"></param>
    ///<param name="org"></param>
    ///<param name="stdLat1"></param>
    ///<param name="stdLat2"></param>
    public LambertProjection(double a, double b, Geographics org,
double stdLat1, double stdLat2)
            :base(a, b)
    {
        _Org = org;
        _stdLat1 = stdLat1;
        _stdLat2 = stdLat2;
    }

    public Geographics XYtoGeographics(double X, double Y)
    {
        double B0 = _Org.B;
        double L0 = _Org.L;
```

```
        //计算投影系数
        double m1 = Math.Cos(_stdLat1) /Math.Sqrt(1 - e1 * e1 *
Math.Sin(_stdLat1) * Math.Sin(_stdLat1));
        double m2 = Math.Cos(_stdLat2) /Math.Sqrt(1 - e1 * e1 *
Math.Sin(_stdLat2) * Math.Sin(_stdLat2));

        double t0 = Math.Tan(Math.PI /4 - B0 /2) /Math.Pow((1 - e1
* Math.Sin(B0)) /(1 + e1 * Math.Sin(B0)), e1 /2);
        double t1 = Math.Tan(Math.PI /4 - _stdLat1 /2) /Math.Pow
((1 - e1 * Math.Sin(_stdLat1)) /(1 + e1 * Math.Sin(_stdLat1)), e1 /
2);
        double t2 = Math.Tan(Math.PI /4 - _stdLat2 /2) /Math.Pow
((1 - e1 * Math.Sin(_stdLat2)) /(1 + e1 * Math.Sin(_stdLat2)), e1 /
2);

        double n = Math.Log(m1 /m2) /Math.Log(t1 /t2);
        double F = m1 /(n * Math.Pow(t1, n));

        //计算 r 值,r0 为原点的 r 值
        double r0 = this.a * F * Math.Pow(t0, n);
        double r = Math.Sqrt(Y * Y + (r0 - X) * (r0 - X)) *
Math.Sign(n);

        //计算经纬度
        double t = Math.Pow(r /(this.a * F), 1 /n);
        double angle = Math.Atan(Y /(r0 - X));

        double lat = getLatitude(t);
        double lng = angle /n + L0;

        return new Geographics(lat, lng);
    }

    //迭代计算纬度值
    private double getLatitude(double t)
    {
        //考虑到中国处于北纬 20~45 度之间,故将初始值定为 20
        double lat = 20 * Math.PI /180;
```

```
            double i = 0;
            do
            {
                double qt = (1 - e1 * Math.Sin(lat)) /(1 + e1 * Math.Sin
(lat));
                double latTemp = Math.PI/2 - 2 * Math.Atan(t * Math.Pow
(qt, e1 /2));
                i = Math.Abs(latTemp - lat);
                lat = latTemp;
            } while (i > 5e-13);

            return lat;
        }
        public Vector2d BLtoVector2d(double B, double L)
        {
            double B0 = _Org.B;
            double L0 = _Org.L;

            //计算投影系数
            double m1 = Math.Cos(_stdLat1) /Math.Sqrt(1 - e1 * e1 *
Math.Sin(_stdLat1) * Math.Sin(_stdLat1));
            double m2 = Math.Cos(_stdLat2) /Math.Sqrt(1 - e1 * e1 *
Math.Sin(_stdLat2) * Math.Sin(_stdLat2));

            double t = Math.Tan(Math.PI /4 - B /2) /Math.Pow((1 - e1 *
Math.Sin(B)) /(1 + e1 * Math.Sin(B)), e1 /2);
            double t0 = Math.Tan(Math.PI /4 - B0 /2) /Math.Pow((1 - e1
* Math.Sin(B0)) /(1 + e1 * Math.Sin(B0)), e1 /2);
            double t1 = Math.Tan(Math.PI /4 - _stdLat1 /2) /Math.Pow
((1 - e1 * Math.Sin(_stdLat1)) /(1 + e1 * Math.Sin(_stdLat1)), e1 /
2);
            double t2 = Math.Tan(Math.PI /4 - _stdLat2 /2) /Math.Pow
((1 - e1 * Math.Sin(_stdLat2)) /(1 + e1 * Math.Sin(_stdLat2)), e1 /
2);

            double n = Math.Log(m1 /m2) /Math.Log(t1 /t2);
            double F = m1 /(n * Math.Pow(t1, n));
```

```
        //计算r值,r0为原点的r值:
        double r = this.a * F * Math.Pow(t, n);
        double r0 = this.a * F * Math.Pow(t0, n);
        double angle = n * (L - L0);

        //计算坐标
        double x = r0 - r * Math.Cos(angle);
        double y = r * Math.Sin(angle);

        return new Vector2d(x, y);
    }
}
```

## 五、功能调用

在 Projection 窗体上创建两个菜单项(Gauss Projection/Lambert Projection)或按钮。然后在 Click 响应函数里,启动 GaussFrm/LambertFrm 即可。

## 六、应上交的成果资料(每人一份)

①各计算公式和编程思想;
②可执行程序与源程序;
③计算结果。

# 实验二　ArcGIS Engine 桌面应用程序框架设计

## 一、目的与要求

①熟悉使用 Visual Studio 2015 开发环境的基本功能；

②熟悉使用 C#进行组件式开发的基本原理和方法(事件驱动机制等)；

③熟悉使用 ArcEngine 中三个框架组件：MapControl、ToolbarControl、TOCControl 构成 GIS 应用程序的基本过程；

④掌握通过菜单添加各类 GIS 功能的基本方法；

⑤掌握通过状态栏显示图形信息的基本方法；

⑥掌握 ArcGIS Engine License 配置方法。

## 二、实验原理

ArcEngine 提供了一系列具有用户界面的组件(简称控件)，包括：

①MapControl：地图控件，显示地图；

②TOCControl：内容列表控件，列表显示已加载的数据层；

③ToolbarControl：工具条控件，是命令控件的驻留容器；

④LayeroutControl：制图布局控件；

⑤ReaderControl；只读控件，等等。

其中，ToolbarControl 和 TOCControl 与 MapControl 协同工作(称之为伙伴控件)，构成了 ArcEngine 应用程序的基本开发框架。

伙伴控件之间通过绑定(Buddy)关联在一起，在设计时可通过 TOCControl 或 ToolbarControl 属性页设置绑定的 MapControl 对象，或在驻留容器被显示时，用 SetBuddyControl 方法通过编程设置绑定的 MapControl 对象。

此外，ArcEngine 提供了许多控件命令，ToolbarControl 可以驻留操作其“伙伴控件”(MapControl 对象)的命令、工具和菜单。开发人员也可以通过创建自定义命令、工具和菜单来扩展 ArcEngine 提供的命令集，这些都可添加到 ToolbarControl 容器中。

## 三、实验环境

①开发环境：Visual Studio 2015 + ArcGIS Engine 10.5；

②开发语言：C#；
③实验数据：... \\ Data \\ 制图数据 \\ ：
- ◆ 地级城市驻地 . shp；
- ◆ 国界线 . shp；
- ◆ 省级行政区 . shp

……

## 四、内容与步骤

本实验使用 MapControl、ToolbarControl、TOCControl 三个控件搭建基本的桌面 GIS 应用程序框架。最终效果预览如图 2-1 所示。

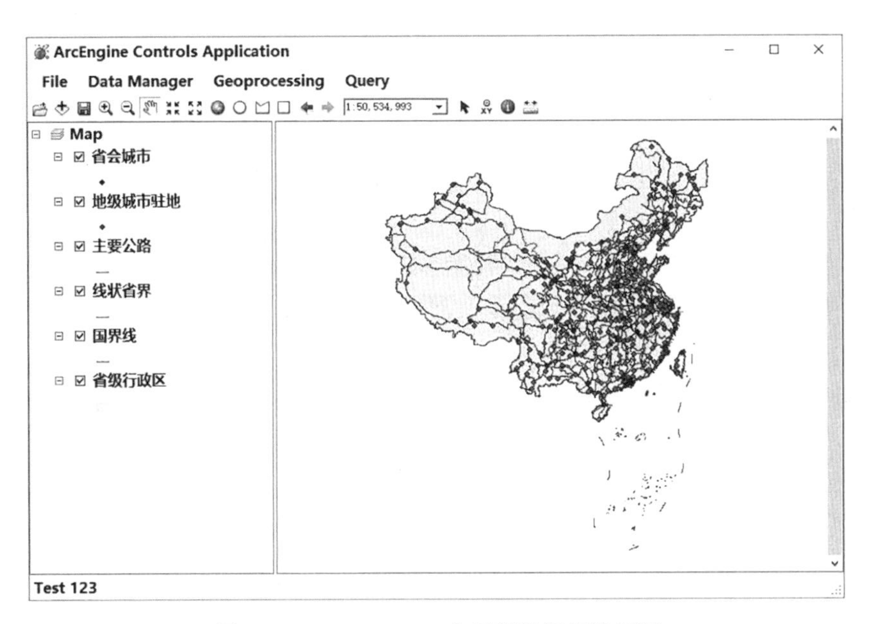


图 2-1　ArcGIS Engine 应用程序框架效果图

### 1. 框架设计

**(1) 新建项目**

启动 VS 2015，选择【文件】→【新建】→【项目】，在项目类型中选择【已安装】模板→【Visual C#】→【ArcGIS】→“Extending ArcObjects 分类目录”，再选择 MapControl Application 模板。如图 2-2 所示，指定项目存放位置(如：C：\ Users)，输入项目名称(默认为“MapControlApplicaton1”)。

点击【确定】，生成应用程序框架，主界面如图 2-3 所示。

在主窗体(MainForm)上排放 MapControl、ToolbarControl、TOCControl 三个控件，还有一个 LicenseControl 控件起授权作用，运行时不可见。

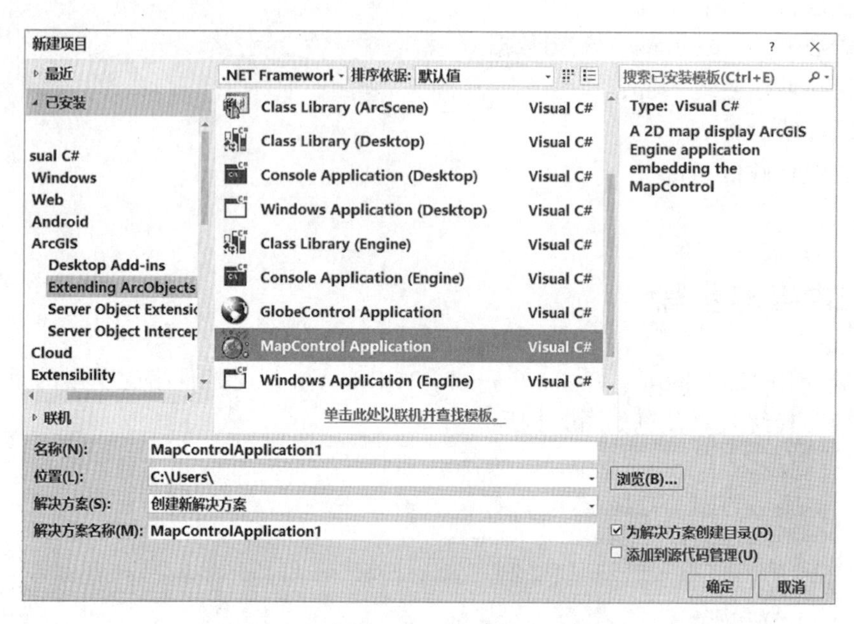


图 2-2　新建 MapControl 应用程序向导

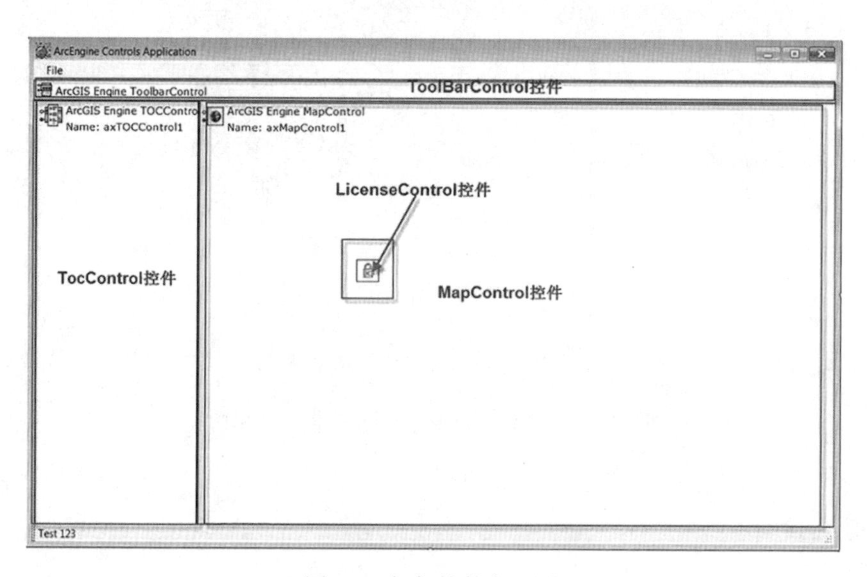


图 2-3　框架控件布置图

**(2)控件绑定**

分别右击 ToolbarControl、TOCControl 控件，打开【属性】对话框，查看 Buddy 下拉框，发现已将 Buddy 属性设置为“axMapControl1(模板自动设置)”，如图 2-4 所示。这样，工具条和图层列表控件就与地图控件关联了。

**(3)添加工具**

右击 ToolbarControl，选择【属性】→【Items】，点击【Add】，出现【Controls Commands】

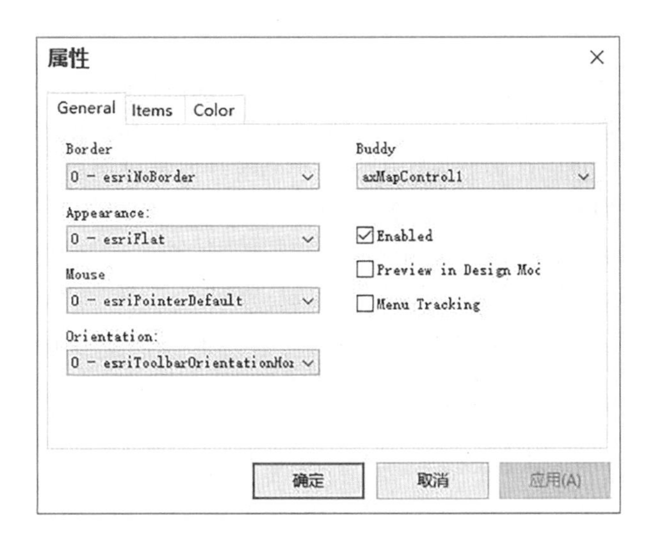


图 2-4　ToolbarControl 属性设置对话框

对话框，如图 2-5 所示，在【Commands】选项卡选中一分类(如“Graphic Element”)，双击右侧列表中的命令(如“New Circle”)或将其拖动到 Items 容器中某个位置，该命令即添加到工具条。

图 2-5　Controls Commands 对话框

ArcGIS Engine 已有 100 多个命令供使用，模板程序已在 ToolbarControl 上放置了一些常用工具(如地图浏览、添加数据等)，其他常见的工具有：Map Navigation 中的导航工具，Map Inquiry 中的查询工具，Feature Selection 中的选择工具，在【Toolsets】选项卡中，Editing 编辑工具也十分有用，你可以根据需要酌情添加。

## 2. MainForm 类说明

MainForm 主窗口类定义了 IMapControl3 类型私有成员：m_mapControl，并在 Load 事件响应函数里赋值。这个十分重要，因为该接口是 axMapControl 负责管理地图(Map)的成员，通过该接口可取得地图的各数据图层，几乎所有功能都可用到(这也就是框架模板专门添加这个私有成员的原因)，代码如下：

```
#region class private members
private IMapControl3 m_mapControl = null;
private string m_mapDocumentName = string.Empty;
#endregion

#region class constructor
public MainForm()
{
  InitializeComponent();
}
#endregion

private void MainForm_Load(object sender, EventArgs e)
{
  //get the MapControl
  m_mapControl = (IMapControl3)axMapControl1.Object;

  //disable the Save menu (since there is no document yet)
  menuSaveDoc.Enabled = false;
}
```

## 3. 添加图层操作菜单(浮动菜单)

在 AE 开发中，右键菜单有两种实现方式：一是使用 VS 带的 ContextMenuStrip 控件，二是用 AE 封装的 IToolbarMenu 接口，本书采用前者实现方法。

**(1)添加菜单及菜单项**

展开工具箱的菜单工具栏，将 ContextMenuStrip 拖入 MainForm 设计视图中，生成名为 contextMenuStrip1 的对象，在属性页中将名称改为"contextMenuTOCLyr"。

上面创建的菜单可认为是菜单容器，里面什么都没有，具体的命令或工具作为菜单项添加到菜单容器里才能工作。在菜单属性 Items 中添加三个菜单项：

①RemoveLayer(移除图层)；

②ZoomToLayer(放大至整个图层)；

③OpenAttributeTable(打开数据表)。

一般情况下，启动程序就要完成菜单项的添加，故此工作在 MainForm_Load 函数中完成。

**(2)添加菜单启动方法**

找到图层列表控件 axTOCControl1 属性页上的 MouseDown 事件，双击响应函数窗格：将在 MainForm 文件里生成响应函数(注意：该操作同时在 MainForm. Designer 文件里添加事件订阅语句)。

补充响应函数代码：

①为 Mainform 类添加 ILayer 私有成员：m_tocRightLayer，用于记录鼠标点击的图层；

②在响应函数中，使用 TOCControl 的 HitTest 函数测试鼠标点击目标的类型，结果为 itemType（esriTOCControlItem 枚举类型)，同时获得 layer(ILayer 类型变量)等；如果鼠标点击在某图层目标上，此时 layer 被赋值(否则为 null)；

③如果 itemType 的值是 esriTOCControlItemLayer，则用 m_tocRightLayer 记录图层对象，然后显示 contextMenuTOCLyr 菜单。

此处：if（e. button = =2）是判断鼠标是否为右键，若为假则返回。

代码如下：

```
    private ILayer m_tocRightLayer = null;
    private void axTOCControl1_OnMouseDown (object sender,
                                    ITOCControlEvents_OnMouseDownEvent e)
    {
        esriTOCControlItem itemType = esriTOCControlItem.
esriTOCControlItemNone;
        IBasicMap basicMap = null;
        ILayer layer = null;
        object unk = null;
        object data = null;
        axTOCControl1.HitTest (e.x, e.y, ref itemType, ref basicMap,
ref layer, ref unk, ref data);
        if (e.button == 2)
        {
            switch (itemType)
            {
                case esriTOCControlItem.esriTOCControlItemLayer:
                    this.m_tocRightLayer = layer;
                    this.contextMenuTOCFeatureLyr.Show(this.
axTOCControl1, e.x, e.y);
                    break;
               case esriTOCControlItem.esriTOCControlItemLegendClass:
                    break;
```

```
            case esriTOCControlItem.esriTOCControlItemMap:
                break;
        }
    }
}
```

**(3)添加 RemoveLayer、ZoomToLayer、OpenAttributeTable 菜单项 Click 响应函数**

代码如下:

```
    private void removeLayerToolStripMenuItem_Click(object sender,
EventArgs e)
    {
        m_mapControl.Map.DeleteLayer(m_tocRightLayer);
    }

    private void zoomToLayerToolStripMenuItem_Click(object sender,
EventArgs e)
    {
    m_mapControl.Extent = m_tocRightLayer.AreaOfInterest;
    }

    private void openAttributeTableToolStripMenuItem_Click(object
sender, EventArgs e)
    {

    }
```

说明：用于响应菜单事件的命令或工具，可以是 AE 内置的也可以是自定义的。AE 内置了许多可以直接调用的常用命令和工具，如 ControlsAddDataCommandClass，在 ESRI. ArcGIS. Controls 命名空间中，可以在对象浏览器中查看。

## 五、License 配置

配置 License 的简单方法是通过设置 LicenseControl 控件来完成，操作方法如下：右键点击【LicenseControl】，点击【属性】菜单。浏览弹出的 LicenseControl 设置对话框，如图 2-6 所示，在左侧“Products”中选中某个授权级别(如图 2-6 中“ArcGIS Engine”已选中，ArcInfo 授权级别最高，但应注意不要多选)，如果需要其他扩展模块的许可，可以在右侧选中对应的复选框，点击【确定】按钮。

LicenseControl 配置有可能出现有些功能授权不正常的现象，ESRI 推荐在运行时配置 License，这也是本书推荐的做法。

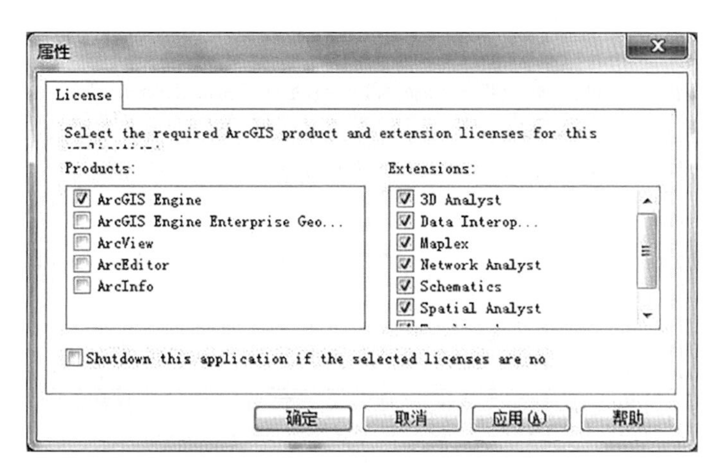


图 2-6　LicenseControl 设置对话框

具体做法是修改 Program 类的 Main( )函数，代码如下：

```
static void Main()
{
    if (! RuntimeManager.Bind(ProductCode.Engine))
    {
        if (! RuntimeManager.Bind(ProductCode.Desktop))
        {
            MessageBox.Show( "Unable to bind to ArcGIS runtime.
Application will be shut down.");
            return;
        }
    }
    LicenseInitializer aoLicenseInitializer = new License
Initializer();
if (! aoLicenseInitializer.InitializeApplication(
new esriLicenseProductCode[]
    { esriLicenseProductCode.esriLicenseProductCodeAdvanced,
esriLicenseProductCode.esriLicenseProductCodeEngineGeoDB },
new esriLicenseExtensionCode[]
{ esriLicenseExtensionCode.esriLicenseExtensionCode
DataInteroperability,
esriLicenseExtensionCode.esriLicenseExtensionCodeSpatialAnalyst,
esriLicenseExtensionCode.esriLicenseExtensionCode3DAnalyst,
esriLicenseExtensionCode.esriLicenseExtensionCodeNetwork} )
```

```
)
    {
        System.Windows.Forms.MessageBox.Show("This application could not initialize with the correct ArcGIS license and will shutdown. LicenseMessage: " + aoLicenseInitializer.LicenseMessage());
        aoLicenseInitializer.ShutdownApplication();
        return;
    }

    Application.EnableVisualStyles();
    Application.SetCompatibleTextRenderingDefault(false);
    Application.Run(new MainForm());

    aoLicenseInitializer.ShutdownApplication();

    //释放 Com 资源
    ESRI.ArcGIS.ADF.COMSupport.AOUninitialize.Shutdown();
}
```

这里用到 LicenseInitializer 密封类，ERSI 给出了详细的源代码，见附录 1。

## 六、编译测试

①单击 F5 键(或点击【调试】菜单→【开始调试】)编译运行；

②点击【AddData】工具添加数据：... \\ Data \\ 制图数据 \\ ：

- 地级城市驻地 . shp
- 国界线 . shp
- 省级行政区 . shp

……

可看到前文图 2-1 的效果图，至此桌面 GIS 应用程序基本框架已经搭建好了。

## 七、思考与练习

①什么是伙伴控件？使用编程方式设置 TOCControl 对象的绑定伙伴控件。

②试手动建立一个与 ArcGIS Engine 模板类似的应用程序框架(参考附录 3)。

# 实验三　Command 命令按钮制作

## 一、目的与要求

①能够使用 C# 建立自定义 GIS 分析组件；
②熟悉用 ArcEngine BaseCommand 模板制作自定义 Command 类型的方法；
③熟悉 ToolBarControl 集成 Command 的基本方法。

## 二、实验原理

ArcEngine 已为用户预定义了上百个 Command 类(AddData、ZoomIn、ZoomOut、Identity 等)，这些类实质上是具有界面的按钮，都可在可视化设计界面拖入 ToolBarControl 驻留容器中，或使用编码添加到 ToolBarControl 中。

这些类都是 BaseCommand 派生类，主要的两个虚函数是：

①OnCreate(object hook)通过传入 hook 钩子对象(通常是 IMapControl3)，初始化命令按钮，其作用相当于 Form 的 Load 事件响应函数；

②OnClick()提供点击命令按钮的事件响应。

开发者可利用 ArcEngine BaseCommand 模板，快速产生一个 BaseCommand 派生类框架，然后做两件事，即实现了新派生类：

①对 BaseCommand 基类表示命令类别，命令名称等属性赋值；

②实现重载 OnClick()。

## 三、实验环境

①开发环境：Visual Studio 2015 + ArcGIS Engine 10.5；
②开发语言：C#；
③实验数据：..\\ Data \\ 制图数据 \\ ：

◆ 地级城市驻地.shp
◆ 国界线.shp
◆ 省级行政区.shp
……

# 四、内容与步骤

## 1. 建立“功能扩展”组件

### (1)新建项目

在 VS 2015 中执行菜单【new】→【Project】，在弹出新建项目对话框，如图 3-1 所示，从模板目录中选择【Visual C#】→【类库】，命名为“BufferAnalyst”。

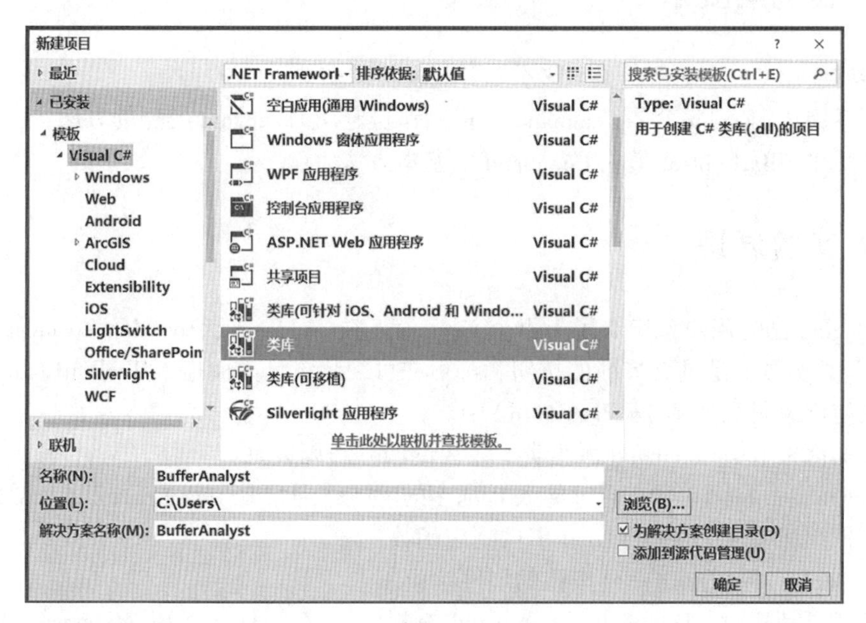


图 3-1　新建类库对话框

### (2)添加缓冲分析对话框类 BufferDlg. cs

BufferDlg. Text 属性设置为 Buffer，修改构造函数如下，对话框中暂时没有实质内容：

```
public  BufferDlg( IMapControl3 mapControl )  {   }
```

### (3)添加缓冲区命令类 BufferSelectedLayerCmd

BufferSelectedLayerCmd 类必须从 AE 中 Base Command 继承，在添加新项时，选择【ArcGIS】→【Extending ArcObjects】目录下的 Base Command 模板，如图 3-2 所示。

```
[Guid("724a2b59-d744-4278-a22e-f513d214d35e")]
[ClassInterface(ClassInterfaceType.None)]
[ProgId("bufferLayer.BufferSelectedLayerCmd")]
public sealed class BufferSelectedLayerCmd : BaseCommand
{
    #region COM Registration Function(s)
```

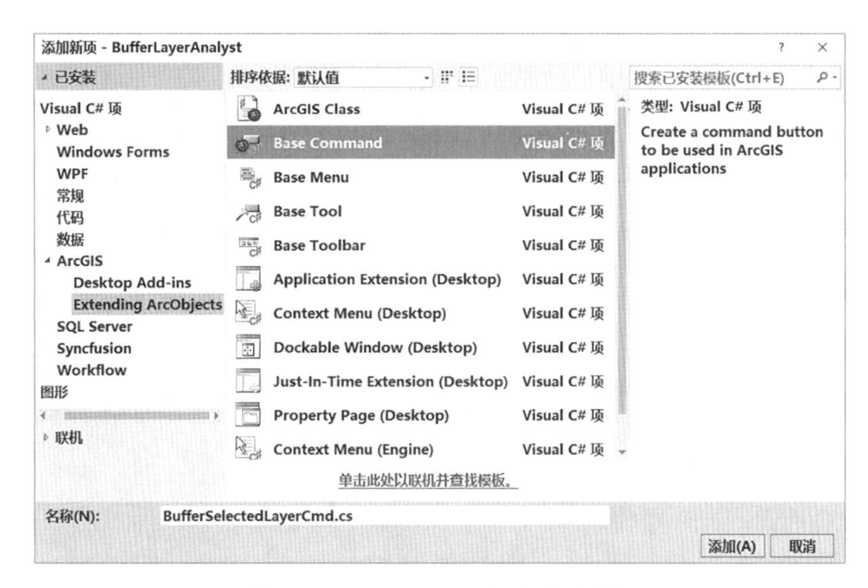


图 3-2　Base Command 新建项对话框

```
[ComRegisterFunction()]
[ComVisible(false)]
static void RegisterFunction(Type registerType)
{
    //Required for ArcGIS Component Category Registrar support
    ArcGISCategoryRegistration(registerType);

    //TODO: Add any COM registration code here
}

[ComUnregisterFunction()]
[ComVisible(false)]
static void UnregisterFunction(Type registerType)
{
    //Required for ArcGIS Component Category Registrar support
    ArcGISCategoryUnregistration(registerType);

    //TODO: Add any COM unregistration code here
}

#region ArcGIS Component Category Registrar generated code
```

```
        ///<summary>
        ///Required method for ArcGIS Component Category
registration -
        /// Do not modify the contents of this method with the code
editor.
        ///</summary>
          private static void ArcGISCategoryRegistration (Type
registerType)
        {
            string regKey = string.Format("HKEY_CLASSES_ROOT\CLSID\
{{{0}}}", registerType.GUID);
            MxCommands.Register(regKey);
            ControlsCommands.Register(regKey);
        }
        ///<summary>
        ///Required method for ArcGIS Component Category
unregistration -
        /// Do not modify the contents of this method with the code
editor.
        ///</summary>
          private static void ArcGISCategoryUnregistration (Type
registerType)
        {
            string regKey = string.Format("HKEY_CLASSES_ROOT\CLSID\
{{{0}}}", registerType.GUID);
            MxCommands.Unregister(regKey);
            ControlsCommands.Unregister(regKey);
        }

        #endregion
        #endregion

        private IHookHelper m_hookHelper = null;
        public BufferSelectedLayerCmd()
        {
            base.m_category = "自定义工具集";
            base.m_caption = "缓冲区分析";
            base.m_message = "缓冲区分析";
```

```
            base.m_toolTip = "缓冲区分析";
            base.m_name = "BufferSelectedLayerCmd";

            try
            {
                string bitmapResourceName = GetType().Name + ".bmp";
                base.m_bitmap = new Bitmap(GetType(), bitmapResource
Name);
            }
            catch (Exception ex)
            {
                    System.Diagnostics.Trace.WriteLine ( ex.Message, "
Invalid Bitmap");
            }
        }

        #region Overriden Class Methods
        ///<summary>
        ///Occurs when this command is created
        ///</summary>
        ///<param name="hook">Instance of the application</param>
        public override void OnCreate(object hook)
        {
            if (hook == null)
                return;

            if (m_hookHelper == null)
                m_hookHelper = new HookHelperClass();

            m_hookHelper.Hook = hook;
        }
        ///<summary>
        ///Occurs when this command is clicked
        ///</summary>
        public override void OnClick()
        {
            if (null == m_hookHelper)
                return;
```

```
            if (m_hookHelper.FocusMap.LayerCount > 0)
            {
                bufferDlg bufferDlg = new bufferDlg(m_hookHelper.
FocusMap);
                bufferDlg.Show();
            }
        }
        #endregion
    }
```

注意：必须重写 OnCreate(object hook)、OnClick()函数以实现启动 BufferDlg 的功能，在构造函数 BufferSelectedLayerCmd()中修改成员变量(base. m_category、base. m_caption 等)，使类库功能名称正确。

2. 集成缓冲区分析工具

①打开主程序工程 MainForm 设计界面；

②在 ToolBarControl 上单击右键，选择【属性】，弹出属性设置对话框；

③在属性设置对话框上选择条目，点击【添加】；弹出命令控件对话框(Controls Commands)；

④在命令类中选择“自定义工具集”，选择“缓冲区分析”，可通过双击或者拖放到工具条 Items 容器上；

⑤如果在 Controls Commands 找不到(可能因为系统注册问题)，可在 MainFrm 的 Load 事件响应函数中添加如下代码：

```
int index = axToolbarControl1.Count;  //工具条中命令个数
axToolbarControl1.AddItem ( newBufferSelectedLayerCmd ( ), - 1,
index, false, -1, esriCommandStyles.esriCommandStyleIconOnly);
```

## 五、编译测试

点击工具条上的【缓冲区分析】按钮，会弹出一个空白对话框。

# 实验四　简单符号化

## 一、目的与要求

①熟练掌握 ISymbol 及 IMarkerSymbol、ILineSymbol、IFillSymbol 符号化接口，及其实现这些接口的组件类的基本用法；

②熟练掌握为图层配置符号的过程与步骤；

③熟练掌握 Windows 对话框 ColorDialog 的用法。

## 二、实验原理

图层符号化，就是使用适当的符号(形态、大小、颜色)，将图层中各要素形象直观地表达出来的过程(修饰要素也是用符号)。在 ArcGTS Engine 中，ISymbol 接口定义了所有符号样式的基本特征(即所有符号化类都继承了这个接口)，并根据点、线、面三种要素几何类型分别定义了 IMarkerSymbol、ILineSymbol、IFillSymbol 三种符号样式，另外，还有两种特殊的符号：TextSymbol(文字标注)、ChartSymbol(图表制图)。为满足各种符号化要求，ArcGTS Engine 提供数十种符号化类供用户使用，具体可参看 ISymbol 接口帮助。

为地图中要素配置指定的符号样式基本思路如下：

①通过 IMapControl3 接口的 get_Layer( )方法获取要素所在图层。

②依据要素的几何类型，创建相应类型的符号：例如，点类型图层，需要创建实现 IMarkerSymbol 的派生类(例如，SimpleMarkerSymbolClass)对象。

③用上一步创建的符号创建一个渲染器(例如，ISimpleRenderer)。

④为图层配置渲染器：先将要素图层转换为 IGeoFeatureLayer 接口(该接口负责图层符号化显示)，然后对 IGeoFeatureLayer 的 Renderer 属性赋值。

⑤刷新：地图刷新使用 Refresh( )，TOCControl 图例更新使用 Update( )。

## 三、实验环境

①开发环境：Visual Studio 2015 + ArcGIS Engine 10. 5；

②开发语言：C#；

③实验数据：.. \\ Data \\ 制图数据 \\ (全国政区人口数据 shapefile)：

◆　地级城市驻地 . shp；

- ◆ 国界线 . shp;
- ◆ 省级行政区 . shp

……

## 四、内容与步骤

本实验实现符号设置基本功能：用户右击 TOCControl 控件中图层，在弹出的菜单中选中符号化菜单项，激活符号化对话框，用户可以调整符号的颜色、线宽、角度等参数。

1. 符号化窗体设计

**(1)新建 Windows 窗体**

新建 Windows 窗体，并命名为“SymbologyFrm”，修改窗体的 Text 属性为“符号化”。并添加 Button、Label、NumericUpDown、ColorDialog 控件。控件布局如图 4-1 所示。

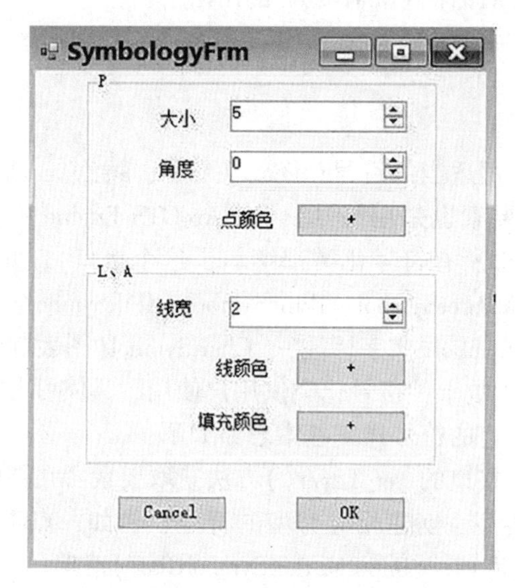


图 4-1 SymbologyFrm 控件布局

**(2)设置控件属性**

设置控件的相关属性，见表 4-1(空白则表示不用修改)。

表 4-1 **SymbologyFrm 窗体控件命名表**

| 控件 | Name 属性 | 含义 | 备注 |
| --- | --- | --- | --- |
| NumericUpDown | nudSize | 点符号大小 | 适用点符号 |
| NumericUpDown | nudAngle | 点符号角度 | 适用点符号 |
| Button | btnMarkerColor | 点符号颜色按钮 | 适用点符号 |

续表

| 控件 | Name 属性 | 含义 | 备注 |
| --- | --- | --- | --- |
| NumericUpDown | nudWidth | 线符号线宽 | 适用线、面状符号 |
| Button | btnOutLineColor | 线颜色按钮 | 适用线、面状符号 |
| Button | btnFillColor | 填充颜色按钮 | 仅适用面状符号 |
| Button | btnOK | 确定 | DialogResult 属性设为 OK |
| Button | btnCancel | 取消 | |
| ColorDialog | colorDialog1 | 颜色对话框 | |

注：三个颜色选择按钮，其选定颜色保存为按钮背景颜色。

**(3)添加引用**

在解决方案资源管理器中添加 ArcGIS Engine 的 ESRI. ArcGIS. Geodatabase 等引用，在 SymbologyFrm. cs 文件中添加如下 Using 指令：

```
usingESRI.ArcGIS.Carto;
usingESRI.ArcGIS.Display;
usingESRI.ArcGIS.esriSystem;
usingESRI.ArcGIS.SystemUI;
usingESRI.ArcGIS.Controls;
usingESRI.ArcGIS.Geodatabase;
```

**(4)添加 SymbologyFrm 的全局变量**

```
privateIFeatureLayer m_pLayer;
privateISymbol m_pSymbol;
```

**(5)添加 ISymbol 公有属性**

添加 ISymbol 公有属性，供其他类使用。

```
public ISymbol pSymbol;
```

**(6)添加 SymbologyFrm 功能函数**

①添加颜色选择按钮 Click 事件响应函数(三个)，为符号提供颜色(点符号颜色、外框线颜色、面填充颜色)选择功能；

②添加确定按钮的 Click 事件响应函数，提供符号创建功能；

代码如下：

```
usingESRI.ArcGIS.Carto;
usingESRI.ArcGIS.Display;
usingESRI.ArcGIS.esriSystem;
usingESRI.ArcGIS.SystemUI;
usingESRI.ArcGIS.Controls;
usingESRI.ArcGIS.Geodatabase;
```

```
public partial class SymbologyFrm : Form
{
    private IFeatureLayer m_pLayer;
private ISymbol m_pSymbol;
//构造函数
    public SymbologyFrm()
    {
        InitializeComponent();
}
//ISymbol 属性
public ISymbol pSymbol
    {
        get {  return m_pSymbol;  }
    }
    //点颜色按钮 Click 响应函数
    private void btnMarkerColor_Click(object sender, EventArgs e)

    //线颜色按钮 Click 响应函数
    private void btnOutLineColor_Click(object sender, EventArgs e)

    //填充颜色按钮 Click 响应函数
    private void btnFillColor_Click(object sender, EventArgs e)

    //确定按钮 Click 响应函数
    private void btnOK_Click(object sender, EventArgs e)

}
```

2. SymbologyFrm 类的实现

**(1)修改 SymbologyFrm 的构造函数**

在 SymbologyFrm 构造函数中添加 IFeatureLayer 参数，用于传入图层接口。代码如下：

```
///<summary>
///构造函数,初始化全局变量
///</summary>
///<param name="tempLayer">图层</param>
publicSymbologyFrm(IFeatureLayer tempLayer)
{
    InitializeComponent();
```

```
        this.m_pLayer = tempLayer;
        this.m_pSymbol = null;
    }
```

**(2)实现颜色选择 Click 响应函数**

通过打开 colorDialog1 颜色对话框选定颜色，结果保存为按钮背景颜色。代码如下：

```
    private void btnMarkerColor_Click(object sender, EventArgs e)
    {
        if (this.colorDialog1.ShowDialog() == DialogResult.OK)
        {
            this.btnMarkerColor.BackColor = colorDialog1.Color;
        }
    }
```

**(3)实现确定按钮 Click 响应函数**

具体步骤如下：

①根据要素图层的几何类型，创建相应类型的符号；

②设置符号的属性；

其中，将.NET 中的 Color 结构转换为 ArcGIS Engine 中的 IColor 接口，用到辅助函数 ColorToIRgbColor(…)，具体代码如下：

```
    /// <summary>
    ///确定按钮
    /// </summary>
    /// <param name="sender"></param>
    /// <param name="e"></param>
    private void btnOK_Click(object sender, EventArgs e)
    {
        switch (this.m_pLayer.FeatureClass.ShapeType)
        {
            case esriGeometryType.esriGeometryPoint:
                IMarkerSymbol pMarkerSymbol =  new SimpleMarker
SymbolClass();;
                //配置点符号大小、角度和颜色
                pMarkerSymbol.Size = (double)this.nudSize.Value;
                pMarkerSymbol.Angle = (double)this.nudAngle.Value;
                pMarkerSymbol.Color = ColorToIRgbColor(this.btnMarker
Color.BackColor);

                m_pSymbol = pMarkerSymbol as ISymbol;
                break;
```

```
            case esriGeometryType.esriGeometryPolyline:
                ILineSymbol pLineSymbol = new SimpleLineSymbolClass();
               //配置线符号线宽和颜色
                pLineSymbol.Width = Convert.ToDouble(this.nudWidth.
Value);
                pLineSymbol.Color = ColorToIRgbColor(this.btnOutline
Color.BackColor);

                m_pSymbol = pLineSymbol as ISymbol;
                break;
            case esriGeometryType.esriGeometryPolygon:
                IFillSymbol pFillSymbol = new SimpleFillSymbolClass();

                //配置轮廓线符号
                ILineSymbol pLSymbol = pFillSymbol.Outline;
                pLSymbol.Color = ColorToIRgbColor(this.btnOutline
Color.BackColor);
                pLSymbol.Width = Convert.ToDouble(this.nudWidth.
Value);

                //配置填充符号轮廓线和填充颜色
                pFillSymbol.Outline = pLSymbol;
                pFillSymbol.Color = ColorToIRgbColor(this.btnFill
Color.BackColor);

                m_pSymbol = pFillSymbol as ISymbol;
                break;
        }
    }
```

**(4)实现辅助函数 ColorToIRgbColor**

具体实现代码如下:

```
    ///<summary>
    ///将.NET中的Color结构转换为ArcGIS Engine中的IColor接口
    ///</summary>
    ///<param name="color">.NET中的System.Drawing.Color结构</
param>
    ///<returns>ARCEngine IColor</returns>
    public IColor ColorToIRgbColor(Color color)
```

```
    {
        IColor pColor = new RgbColorClass();
        pColor.RGB = color.B * 65536 + color.G * 256 + color.R;
        return pColor;
    }
```

### 3. 调用符号化窗体

在图层操作浮动菜单上添加一菜单项(命名为 Single Symbolize)，创建并修改 Click 事件响应函数，代码如下：

```
    private void singleSymbolizeToolStripMenuItem_Click(object sender,
EventArgs e)
    {
        //创建符号化实例
         SymbologyFrm SymbolFrm  = new SymbologyFrm(this.m_tocRight
Layer as IFeatureLayer);
        if (SymbolFrm.ShowDialog() == DialogResult.OK)
        {
            //局部更新主 Map 控件
            m_mapControl.ActiveView.PartialRefresh(esriViewDrawPhase.
esriViewGeography, null, null);

            //初始化渲染器,并配置符号
            ISimpleRenderer renderer = new SimpleRendererClass();
            renderer.Symbol = SymbolFrm.pSymbol;

            //为图层配置渲染器
             IGeoFeatureLayer geoFeatureLyr = this.m_tocRightLayer as
IGeoFeatureLayer;
            geoFeatureLyr.Renderer = renderer as IFeatureRenderer;

            //更新 Map 控件和图层控件
            this.axMapControl1.ActiveView.Refresh();
            this.axTOCControl1.Update();
        }
    }
```

## 五、编译与测试

①单击 F5 键编译运行；

②加载 . . \\ Data \\ 制图数据 \\ ：

- ◆ 地级城市驻地 . shp；
- ◆ 国界线 . shp；
- ◆ 省级行政区 . shp；

……

③右键选择图层，在浮动菜单中点击【Single Symbolize】菜单，分别为每个图层配置符号的颜色和大小，效果如图 4-2 所示。

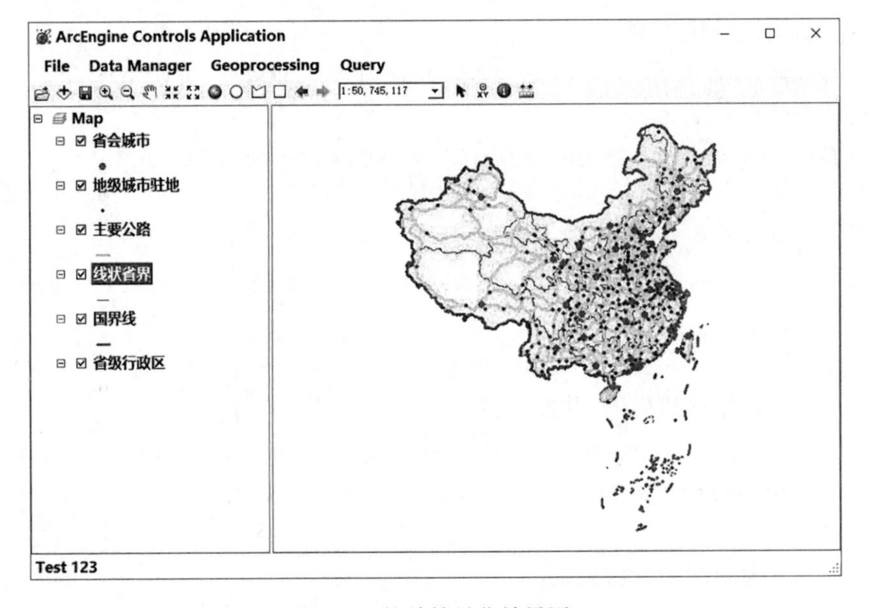


图 4-2　简单符号化效果图

## 六、思考与练习

①修改本章程序，用箭头符号代替简单点符号；

②修改本章程序，用虚线符号代替简单线符号；

③修改本章程序，用斜线填充符号代替简单填充符号。

# 实验五　唯一值符号化

## 一、目的与要求

①熟悉 IUniqueValueRenderer 渲染器接口的用法；
②熟悉 IRandomColorRamp 色带接口的用法；
③掌握唯一值集合获取方法。

## 二、实验原理

ArcGIS Engine 由 IFeatureRender 接口负责要素渲染，其子类包括：简单绘制(SimpleRenderer)；唯一值绘制(UniqueValueRenderer)或多字段唯一值绘制；点密度或多字段点密度绘制(DotDensityRenderer)；数据分级绘制(ClassBreaksRenderer)；饼图或直方图(ChartRenderer)；比例符号渲染(ScaleDependentRenderer)。

唯一值绘制方法是依据要素图层的要素类中的字段值唯一值集合，对每个值分别进行渲染，使用唯一值绘制可以通过颜色区分每个要素。UniqueValueRenderer 对象类实现了 IUniqueValueRenderer 接口，该接口定义了几个重要属性和方法。

①FieldCount：需要显示的唯一分类的字段数目。

②set_Field(int Index, string Field)：设置唯一分类值的字段名。

③FieldDelimiter：多个字段组合时，字段值分格符。

④AddValue (string Value, string Heading, ISymbol Symbol)：添加将(值，符号)对 UniqueValueRenderer 要素类渲染的使用方法：先遍历图层中的所有要素，获取待渲染字段的唯一值集合，然后创建一个 UniqueValueRenderer 对象，使用 AddValue()方法添加“值-符号”对，将唯一值渲染器赋给图层的 Renderer 属性。

## 三、实验环境

①开发环境：Visual Studio 2015 + ArcGIS Engine 10.5；
②开发语言：C#；
③实验数据：.. \\ Data \\ 制图数据 \\ (全国政区人口数据 shapefile)：
- ◆ 地级城市驻地 . shp；
- ◆ 国界线 . shp；

◆ 省级行政区 . shp;

……

## 四、内容与步骤

本实验实现唯一值符号化功能：用户右键单击 TOCControl 控件中图层，在弹出的菜单中选中唯一值符号化(UniqueValueRenderer)菜单项，激活唯一值符号化对话框，用户可选择用于符号化的唯一值字段。

### 1. 符号化窗体设计

**(1)新建 Windows 窗体**

新建 Windows 窗体，并命名为“UniqueValueRendererFrm”，修改窗体的 Text 属性为“唯一值符号化”，并添加 Button、Label、Combox 控件。控件布局如图 5-1 所示。

图 5-1 UniqueValueRendererFrm 控件布局

**(2)设置控件属性**

设置控件的相关属性，见表 5-1(空白则表示不用修改)。

表 5-1 **UniqueValueRendererFrm 控件命名表**

| 控件 | Name 属性 | 含义 | 备注 |
|---|---|---|---|
| Combox | cbxFieldName | 唯一值字段 | 下拉框列举可选字段 |
| Button | btnOK | 确定 | DialogResult 属性设为 OK |
| Button | btnApp | 应用 | 执行核心功能 |

**(3)添加引用**

在解决方案资源管理器中添加 ArcGIS Engine 的 ESRI. ArcGIS. Geodatabase 等引用，在 UniqueValueRendererFrm. . cs 文件中添加如下 Using 指令：

```
using ESRI.ArcGIS.Carto;
using ESRI.ArcGIS.Controls;
```

```
using ESRI.ArcGIS.Display;
using ESRI.ArcGIS.Geodatabase;
using ESRI.ArcGIS.Geometry;
using System;
using System.Collections;
using System.Collections.Generic;
using System.Windows.Forms;
```

**(4)添加 SymbologyFrm 的全局变量**

```
private IMapControl3 _mapControl;
private IFeatureLayer _pFeatLyr;
```

**(5)添加 UniqueValueRendererFrm 事件响应函数、功能函数、辅助函数**

①添加应用按钮 Click 事件响应函数;

②添加确定按钮 Click 事件响应函数;

具体代码如下:

```
public partial class UniqueValueRendererFrm : Form
{
    IMapControl3 _mapControl = null;
    IFeatureLayer _pFeatLyr = null;
    public UniqueValueRendererFrm( )

    //消息响应函数
     private void UniqueValueRendererFrm _ Load ( object sender,
EventArgs e)
    private void btnCancel_Click(object sender, EventArgs e)
    private void btnApp_Click(object sender, EventArgs e)
    private void btnOK_Click(object sender, EventArgs e)

    //核心功能函数
     private void UniqueValueRenderer ( IFeatureLayer pFeatLyr,
string sFieldName)

//辅助函数:获取指定字段唯一值集合
private List < string > GetUniqueValues ( ICursor pCursor, string
sFieldName)
  //辅助函数:创建随机色带
    private IRandomColorRamp CreateColorRamp(int nSize)
//辅助函数:创建指定类型的符号
     private ISymbol CreateSymbol ( esriGeometryType shapeType,
```

```
IColor pUniqueColor)
    }
```

2. UniqueValueRendererFrm 类的实现

**(1)修改构造函数**

构造函数添加 IMapControl3、IFeatureLayer 参数，用于传入图层接口等。代码如下：

```
    /// <summary>
    ///构造函数,初始化全局变量
    /// </summary>
    /// <param name="tempLayer">图层</param>
    public UniqueValueRendererFrm(IMapControl3 mapControl, IFeatureLayer
pFeatureLayer)
    {
        InitializeComponent();
        _pFeatLyr = pFeatureLayer;
        _mapControl = mapControl;
    }
```

**(2)实现 Load 事件响应函数**

对话框加载时，用图层的字段名填充字段名 ComboBox 控件(cbxFieldName)，代码如下：

```
    private void UniqueValueRendererFrm_Load(object sender,
EventArgs e)
    {
        IFields pFields = _pFeatLyr.FeatureClass.Fields;
        this.cbxFieldName.Items.Add("");
        for (int k = 0; k < pFields.FieldCount; k++)
        {
            IField pFd = pFields.get_Field(k);
            this.cbxFieldName.Items.Add(pFd.Name);
        }

        this.cbxFieldName.SelectedIndex = 0;
    }
```

**(3)"应用"按钮 Click 响应函数**

根据选定的字段，调用核心函数 UniqueValueRenderer 为图层配置唯一值渲染器，然后刷新地图。代码如下：

```
    private void btnApp_Click(object sender, EventArgs e)
    {
```

```
        string sFieldName = this.cbxFieldName.SelectedItem.ToString();
        if (sFieldName.Length > 0)
        {
            UniqueValueRenderer(_pFeatLyr, sFieldName);
            _mapControl.ActiveView.Refresh();
        }
    }
```

**(4)实现核心函数：UniqueValueRenderer()**

实现步骤如下：

①获取字段组合唯一值集合；

②创建随机色带；

③初始化唯一值渲染器；

④遍历唯一值集合，为渲染器配置符号；

⑤为图层配置唯一值渲染器。

代码如下：

```
    private void UniqueValueRenderer(IFeatureLayer pFeatLyr, string sFieldName)
    {
        //获取字段组合唯一值集合
        ITable pTable = pFeatLyr.FeatureClass as ITable;
        ICursor pCursor = pTable.Search(null, true);
        List<string> UniqueValueArr = GetUniqueValues(pCursor, sFieldName);

        //创建颜色枚举器:
        int nSize = UniqueValueArr.Count;
        IRandomColorRamp pRandColorRamp = CreateColorRamp((nSize < 100) ? nSize : 100);
        IEnumColors pEnumRamp = pRandColorRamp.Colors;

        //初始化唯一值渲染器
        IUniqueValueRenderer pUVRender = new UniqueValueRendererClass();
        pUVRender.FieldCount = 1;
        pUVRender.set_Field(0, sFieldName);

        //遍历唯一值集合,为渲染器配置符号
```

```
        esriGeometryType shapeType = pFeatLyr.FeatureClass.ShapeType;
        foreach(string sValue in UniqueValueArr)
        {
            //获取下一个颜色值
            IColor pNextUniqueColor = pEnumRamp.Next();
            if (pNextUniqueColor == null)
            {
                pEnumRamp.Reset();
                pNextUniqueColor = pEnumRamp.Next();
            }
            //创建符号;
            ISymbol pSymbol = CreateSymbol(shapeType, pNextUniqueColor);

            //配置唯一值渲染器
            pUVRender.AddValue(sValue, null, pSymbol);
        }

        //为图层配置唯一值渲染器
        IGeoFeatureLayer pGeoFeatureL = pFeatLyr as IGeoFeatureLayer;
        pGeoFeatureL.Renderer = (IFeatureRenderer)pUVRender;
    }
```

**(5)实现辅助函数：GetUniqueValues()**

具体代码如下：

```
    private List<string> GetUniqueValues(ICursor pCursor, string
sFieldName)
    {
        List<string> UniqueValueArr = new List<string>();
        Hashtable hashTable = new Hashtable();

        //字段索引
        int index = pCursor.Fields.FindField(sFieldName);
        IRow pNextRow = null;
        while ((pNextRow = pCursor.NextRow()) != null)
        {
            //计算字段组合值,图例标头;
            object obj = pNextRow.get_Value(index);
            string sValue = obj.ToString();
```

```
            //配置唯一值渲染器:用 Hash 表控制重复:
            if (! hashTable.Contains(sValue))
            {
                UniqueValueArr.Add(sValue);
                hashTable.Add(sValue, sValue);
            }
        }

        return UniqueValueArr;
    }
```

**(6)实现辅助函数：CreateColorRamp()**

具体代码如下：

```
    private IRandomColorRamp CreateColorRamp(int nSize)
    {
         IRandomColorRamp pRandColorRamp = new RandomColorRampClass
();
        {
            pRandColorRamp.StartHue = 0;
            pRandColorRamp.EndHue = 360;

            pRandColorRamp.MinValue = 0;
            pRandColorRamp.MaxValue = 100;

            pRandColorRamp.MinSaturation = 15;
            pRandColorRamp.MaxSaturation = 30;

            pRandColorRamp.Size = nSize;
        }

        bool bSuccess = false;
        pRandColorRamp.CreateRamp(out bSuccess);

        return pRandColorRamp;
    }
```

**(7)实现辅助函数：CreateSymbol()**

具体代码如下：

```
    private ISymbol CreateSymbol(esriGeometryType shapeType, IColor
pUniqueColor)
```

```
    {
        ISymbol pSymbol;
        switch (shapeType)
        {
            case esriGeometryType.esriGeometryPolygon:
                {
                    IFillSymbol pFillSymbol = new SimpleFillSymbolClass();
                    pFillSymbol.Color = pUniqueColor;
                    pSymbol = (pFillSymbol as ISymbol);
                    break;
                }
            case esriGeometryType.esriGeometryPolyline:
                {
                    ISimpleLineSymbol pLineSymbol = new SimpleLine
SymbolClass();
                    pLineSymbol.Color = pUniqueColor;
                    pSymbol = (pLineSymbol as ISymbol);
                   break;
                }
            case esriGeometryType.esriGeometryPoint:
            default:
                {
                    IMarkerSymbol pMarkerSymbol = new SimpleMarker
SymbolClass();
                    pMarkerSymbol.Color = pUniqueColor;
                    pSymbol = (pMarkerSymbol as ISymbol);
                    break;
                }
        }
        return pSymbol;
    }
```

3. 调用唯一值符号化窗体

在图层操作浮动菜单上添加一菜单项(命名为 UniqueValueRenderer)，创建并修改 Click 事件响应函数，代码如下：

```
    private void uniqueValueRendererToolStripMenuItem_Click(object
sender, EventArgs e)
    {
```

```
//创建唯一值符号化实例夹
UniqueValueRendererFrm SymbolFrm =new UniqueValueRendererFrm
(m_mapControl, this.m_tocRightLayer as IFeatureLayer);
if (SymbolFrm.ShowDialog() = = DialogResult.OK)
{
    //更新主 Map 控件和图层控件
    this.axMapControl1.ActiveView.Refresh();
    this.axTOCControl1.Update();
}
}
```

## 五、编译测试

①单击 F5 键编译运行；
②加载数据：.. \\ Data \\ 制图数据 \\ 省级行政区 . shp；
③启动唯一值符号功能，选择唯一值字段 =【NAME】；
④点击【应用】按钮；
按 NAME 唯一值符号化效果如图 5-2 所示。

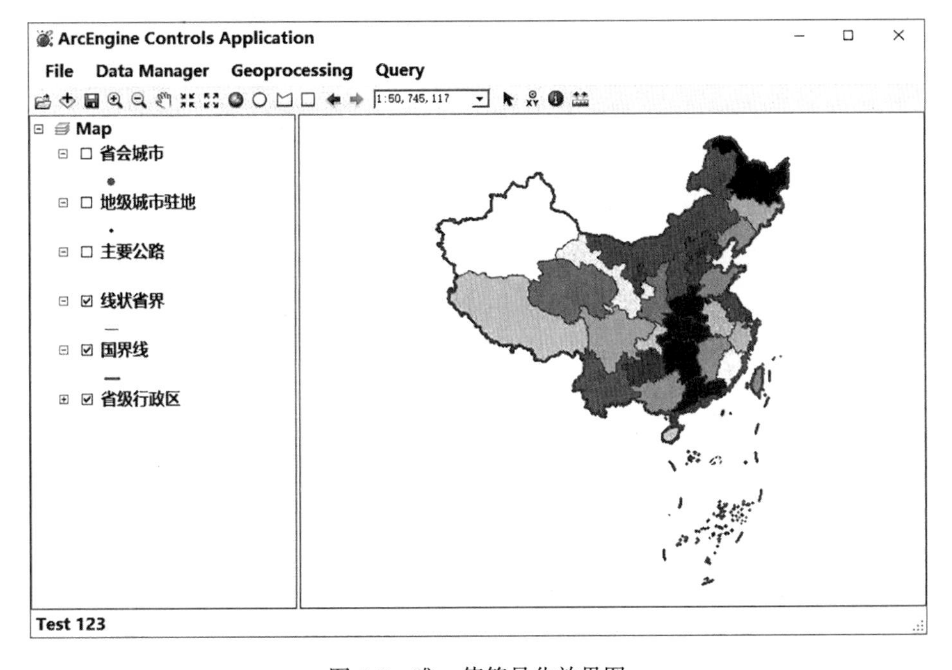


图 5-2　唯一值符号化效果图

## 六、思考与练习

以本实验代码为基础，将功能扩展为支持两个字段组合的唯一值符号化。

# 实验六 栅格数据分级渲染

## 一、目的与要求

①理解属性值分级分类的基本原理；
②熟悉栅格数据符号化的一般过程；
③掌握 IRasterClassifyColorRampRenderer 构建方法。

## 二、实验原理

分级色彩渲染是指按照一定的分级方法将需渲染的属性值分成若干级别，再用不同的颜色来表示。在 ArcG1S Engine 中 IRasterClassifyColorRampRenderer 接口具有栅格数据分级色彩渲染功能，其 Field 属性用于设置分级的字段(对于栅格数据，通常是 Value)，BreakCount 属性用于设置分级的数目，SortClassesAscending 属性用于设置分级后图例是否按升级顺序排列。矢量数据分级色彩渲染由 IClassBreaksRenderer 接口提供，用法类似。

实验代码的实现思路如下：

①对渲染字段划分等级，IClassifyGEN 的子类实现了不同的分级方法(等间距分级、等差分级、等比分级、自然断点法等)，但是 IClassifyGEN 接口需要栅格属性值的频数表。为简化起见，本实验采用人工等间距分类。

②初始化 IRasterClassifyColorRampRenderer 接口，并利用 Symbol 和 Break 属性进行分级颜色设置。该接口由 RasterClassifyColorRampRenderer 类实现。

③为指定图层配置渲染器，类似于矢量图层。

## 三、实验环境

①开发环境：Visual Studio 2015 + ArcGIS Engine 10. 5；
②开发语言：C#；
③实验数据：... \\ Data \\ 栅格数据 \\ Ex9. gdb \\ ：
◆ DEM_CASE1；
◆ Soil_Case1。

## 四、内容与步骤

本实验实现栅格数据符号化基本功能：用户右击 TOCControl 控件中某图层，在弹出的菜单中选中栅格符号化(Raster Renderer)菜单项，激活栅格符号化对话框，用户可选择用于渲染的级数。

### 1. RasterSymbolize 窗体设计

#### (1)添加栅格符号化对话框类

添加栅格符号化对话框类，并命名为“RasterSymbolizeFrm”，修改窗体的 Text 属性为“RasterSymbolize”，并添加 Button、Label、NumericUpdown 等控件。控件布局如图 6-1 所示。

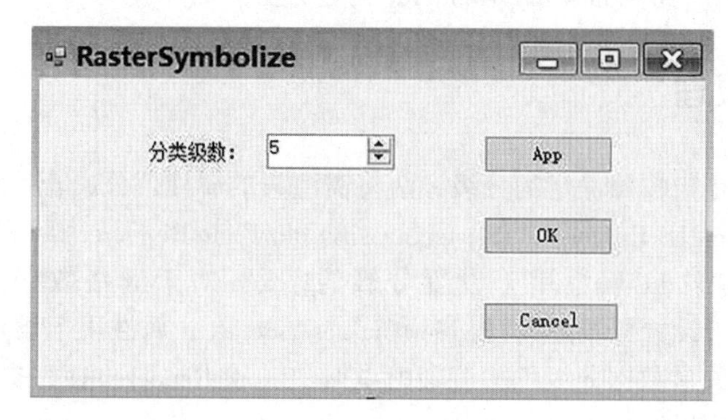


图 6-1　RasterSymbolize 窗体控件布局

#### (2)设置控件属性

设置相应控件的相关属性，见表 6-1。

表 6-1　**RasterSymbolize 窗体控件命名表**

| 控件 | Name 属性 | 含义 |
| --- | --- | --- |
| NumericUpdown | nudClasses | 分级数 |
| Button | btnApp | 应用 |
| Button | btnCancel | 关闭 |
| Button | btnOK | 确定 |

#### (3)添加 RasterSymbolizeFrm 的全局变量

```
private IRasterLayer _rasterLayer = null;
```

#### (4)添加 RasterSymbolizeFrm 事件的响应函数、功能函数、辅助函数

①添加应用按钮 Click 事件响应函数；

②添加确定按钮 Click 事件响应函数；

代码如下：

```
public partial class RasterSymbolize : Form
{
    private IRasterLayer _rasterLayer = null;
    public RasterSymbolize(IRasterLayer rasterLayer)
    {
        InitializeComponent();
        _rasterLayer = rasterLayer;
    }

private void btnOK_Click(object sender, EventArgs e)
private void btnApp_Click(object sender, EventArgs e)
    private void btnCancel_Click(object sender, EventArgs e)

    //构建 RGB 颜色
    private IRgbColor GetRgbColor(int red, int green, int blue)
    //创建起止色带
    private IEnumColors CreateAlgorithmicColorRamp(int Classes
Count)
    //创建分级数据
    private double[] CreateBreakClass(IRasterLayer rasterLayer,
int DesiredClasses)
}
```

## 2. RasterSymbolizeFrm 类的实现

### (1) btnApp 按钮响应函数的实现

btnApp 按钮响应函数可完成栅格数据分级渲染，步骤如下：

第一步，根据界面指定的分级数，创建分级数组，使用辅助函数 CreateBreakClass()；

第二步，创建色带，用到 CreateAlgorithmicColorRamp()函数；

第三步，初始化栅格分级色彩渲染器；

第四步，为渲染器各等级配置颜色；

第五步，为栅格数据图层配置渲染器。

代码如下：

```
private void btnApp_Click(object sender, EventArgs e)
{
    //创建分级数组
    int DesiredClasses = Convert.ToInt32(this.nudClasses.Value);
```

```
        double [ ] dblClasses = CreateBreakClass (_rasterLayer,
DesiredClasses);

        //创建色带
        IEnumColors pEnumColors = default(IEnumColors);
        pEnumColors = CreateAlgorithmicColorRamp(dblClasses.
Length);

        //初始化栅格分级色彩渲染器
        IRasterClassifyColorRampRenderer pClassRenderer = null;
        pClassRenderer = new RasterClassifyColorRampRendererClass();
        {
            pClassRenderer.ClassField = "Value";
            pClassRenderer.ClassCount = dblClasses.Length;
            pClassRenderer.SortClassesAscending = true;
        }

        //为渲染器各等级配置颜色
        for (int i = 0; i < dblClasses.Length; i++)
        {
            IColor pColor = pEnumColors.Next();
            ISimpleFillSymbol pFillSymbol = new SimpleFillSymbol();
            pFillSymbol.Color = pColor;
            pFillSymbol.Style = esriSimpleFillStyle.esriSFSSolid;

            pClassRenderer.set_Symbol(i, pFillSymbol as ISymbol);
            pClassRenderer.set_Break(i, dblClasses[i]);
            pClassRenderer.set_Label(i, dblClasses[i].ToString());
        }

        //指定图层配置渲染器
        _rasterLayer.Renderer = pClassRenderer as IRasterRenderer;
    }

    private void btnCancel_Click(object sender, EventArgs e)
    {
        this.Close();
    }
```

```
private void btnOK_Click(object sender, EventArgs e)
{
    this.Close();
}
```

**(2)辅助函数**

CreateBreakClass()函数创建分级数组，方法是用IRasterBand的ComputeStatsAndHist()函数获得栅格值的统计数据，然后根据最小值和最大值等间距构造分级数组(注意：为简化学习过程，这里仅用简单的分级方法)。具体代码如下：

```
private double[] CreateBreakClass(IRasterLayer rasterLayer, int DesiredClasses)
{
    //获得栅格数据第一波段
    IRasterBandCollection pRsBandCol = rasterLayer.Raster as IRasterBandCollection;
    IRasterBand pRsBand = pRsBandCol.Item(0);

    //获得最大值和最小值,以设置分类级数
    pRsBand.ComputeStatsAndHist();
    IRasterStatistics pRasterStatistic = pRsBand.Statistics;
    double dMaxValue = pRasterStatistic.Maximum;
    double dMinValue = pRasterStatistic.Minimum;

    //构造分级数组
    double BinInterval = Convert.ToDouble((dMaxValue - dMinValue) /DesiredClasses);
    double[] dblValues = new double[DesiredClasses+1];
    for (int i = 0; i < DesiredClasses+1; i++)
    {
        dblValues[i] = i * BinInterval + dMinValue;
    }

    return dblValues;
}

private IEnumColors CreateAlgorithmicColorRamp(int ClassesCount)
{
    //从黄到红的渐变色,设置初始颜色为黄色
```

```
        IHsvColor pFromColor = new HsvColor();
        pFromColor.Hue = 60;            //黄
        pFromColor.Saturation = 100;
        pFromColor.Value = 96;
        //设置结束色为红色
        IHsvColor pToColor = new HsvColor();
        pToColor.Hue = 0;           //红
        pToColor.Saturation = 100;
        pToColor.Value = 96;

        //建立从黄到红的渐变色
        IAlgorithmicColorRamp pRamp = new AlgorithmicColorRamp();
        {
            pRamp.Algorithm = esriColorRampAlgorithm.esriHSVAlgorithm;
            pRamp.FromColor = pFromColor;
            pRamp.ToColor = pToColor;
            pRamp.Size = ClassesCount;
        }

        bool ok = false;
        pRamp.CreateRamp(out ok);

        return pRamp.Colors;
    }
```

此处，CreateAlgorithmicColorRamp()函数建立从黄到红的渐变色带。

GetRgbColor()函数参见实验四《简单符号化》。

### 3. 功能调用

在图层操作浮动菜单上添加一菜单项(命名为 Raster Symbolize)，创建并修改 Click 事件响应函数，代码如下：

```
    private void rasterSymbolizeToolStripMenuItem_Click(object
sender, EventArgs e)
    {
        RasterSymbolize frm = new RasterSymbolize(this.m_tocRight
Layer as IRasterLayer);
        if (frm.ShowDialog() == DialogResult.OK)
        {
            m_mapControl.ActiveView.Refresh();
```

```
        axTOCControl1.Update();
    }
}
```

## 五、编译测试

①单击 F5 键编译运行;
②添加栅格数据:… \\ Data \\ 栅格数据 \\ Ex9. gdb \\ DEM_CASE1;
③点击图层操作浮动菜单【Raster Symbolize】，启动栅格符号化功能，按 9 级符号化;
效果如图 6-2 所示。

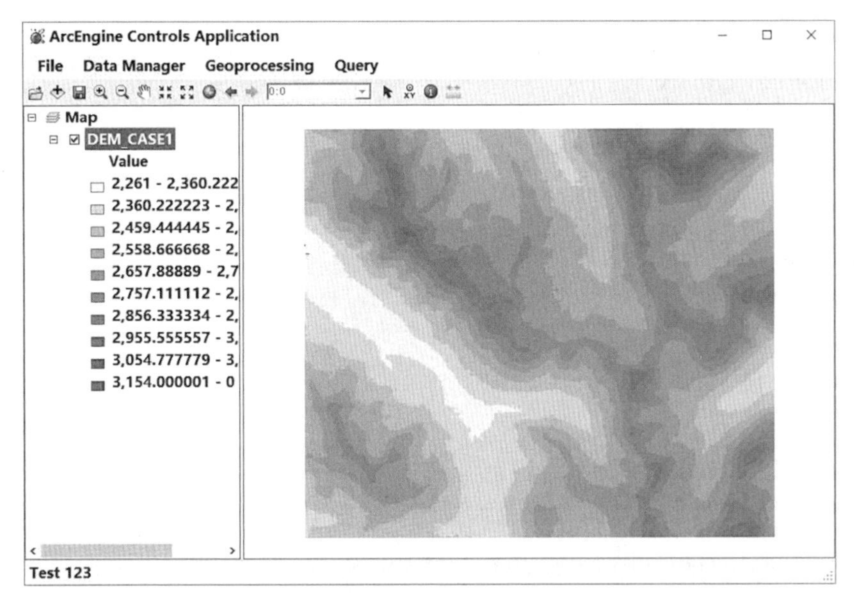


图 6-2　栅格数据分级符号化效果图

## 六、思考与练习

①试实现栅格数据渐变颜色渲染。
②试实现矢量数据分级符号化渲染。

# 实验七　Workspace 加载数据层

## 一、目的与要求

本实验的目的是熟悉以下加载空间数据接口的使用方法：

①常用工作空间工厂：

- ShapefileWorkspaceFactory，针对 Shapefile 文件数据源；
- FileGDBWorkspaceFactory，针对 File Geodatabase 数据源；
- RasterWorkspaceFactory，针对栅格数据源。

②常用工作空间：

- IFeatureWorkspace，针对矢量数据；
- IRasterWorkspace，针对栅格数据。

## 二、实验原理

ArcGIS 采用 Geodatabase 模型管理空间数据。Geodatabase 是 ESRI 在 ArcInfo8 之后引入的一种全新的面向对象的空间数据存储模型，它采用统一的框架面向不同类型的数据源，从而为空间数据管理提供了统一的操作模式。

①Geodatabase 是通过工作空间(Workspace)进行操作的，打开某个数据源的工作空间就意味着建立了该数据源的操作管道。逻辑层面上 Workspace 是包含各种数据集 Dataset(包括空间的和非空间的)的容器，它们都实现了 IDataset 接口，或者说 Workspace 的一切对象都是 IDataset，不管是要素类，还是表格或者栅格数据、要素数据集等。通过 IDataset 的 Type 属性可以判断 IDataset 属于哪一类型数据集。工作空间实现了众多的接口，比如 IWorkspace、IFeatureWorkspace、IRasterWorkspace 等，以满足不同类型数据集操作的需要。

②在物理级别上 Geodatabase 分为三种不同的存储形式：ShapeFile 文件(存放文件的目录起到类似数据库的作用)、本地数据库(个人数据库 mdb，文件数据库 gdb)，以及面向企业的 SDE 数据库。个人数据库依赖于微软的 Access 数据库，个人数据库最大存储量不能超过 2GB。文件数据库是 ESRI 提供的本地数据库，单张表可以存储 1TB。SDE 数据库是通过 Arc SDE 空间数据引擎在大型商业关系型数据库上 Geodatabase 的实现，SDE 数据库功能强大，除支持多用户同时编辑数据之外，还提供了一些其他高级功能，如同步复

制，历史归档等。目前，SDE 支持 5 种数据库（oracle、sql server、db2、infomix、postgresql）。

③Geodatabase 模型中包括含多种类型的工作空间工厂（WorkspaceFactory），例如：ShapeFileWorkspaceFactory、GdbWorkspaceFactory、SdeWorkspaceFactory 等，为不同数据源建立对应的工作空间。这也就说明了为什么 Workspace 是普通类，需通过工作空间工厂（WorkspaceFactary）才能建立。

## 三、实验环境

①开发环境：Visual Studio 2015 + ArcGIS Engine 10.5。
②开发语言：C#。
③实验数据：..\\Data\\制图数据\\（全国政区人口数据 shapefile）：
- 地级城市驻地.shp；
- 国界线.shp；
- 省级行政区.shp；

……

## 四、内容与步骤

**（1）添加启动菜单**

在 Data Manager 主菜单上添加三个菜单项：OpenShpFeatureLayer、OpenRasterLayer、OpenFileGdbLayer，分别用于打开 Shapefile 数据、栅格数据、File Geodatabase 中所有要素类，分别双击菜单项添加三个响应函数：openShpFeatureLayerStripMenuItem_Click（...）；openRasterLayerStripMenuItem_Click（...）；openFileGdbLayerStripMenuItem_Click（...）。

**（2）实现 openShpFeatureLayerStripMenuItem_Click 函数**

openShpFeatureLayerStripMenuItem_Click 函数可实现加载 shp 数据，实现方法如下：
①创建一个 ShapefileWorkspaceFactoryClass 实例；
②使用 IWorkspaceFactory 打开工作空间 IFeatureWorkspace（此处用文件路径名指定）；
③使用 IFeatureWorkspace 打开要素类 IFeatureClass；
④创建要素数据图层 IFeatureLayer，将打开的要素类赋值给 IFeatureLayer 的 FeatureClass 属性；
⑤将要素图层添加到 mapControl 控件。
具体代码如下：

```
private voidopenFeatureLayerStripMenuItem_Click(object sender,
EventArgs e)
{
    string filePath = "C:\\用户目录\\制图数据";
    string fileName = "省级行政区";
```

```
    //1:创建 Shapefile 工作空间工厂
    IWorkspaceFactory pWorkspaceFactory;
    pWorkspaceFactory = new ShapefileWorkspaceFactoryClass();

    //2:打开工作空间
    IFeatureWorkspace pFWorkspace;
    pFWorkspace = pWorkspaceFactory.OpenFromFile(filePath, 0) as
    IFeatureWorkspace;

    //3:打开矢量数据
    IFeatureClass pFeatureClass;
    pFeatureClass = pFWorkspace.OpenFeatureClass(fileName);

    //4:新建要素数据层
    IFeatureLayer pFeatureLayer = new FeatureLayerClass();
    pFeatureLayer.FeatureClass = pFeatureClass;
    pFeatureLayer.Name = fileName;

    //5:将新建层添加到 MapControl
    axMapControl1.AddLayer(pFeatureLayer);
    axMapControl1.Refresh();
}
```

**(3)实现 openRasterLayerStripMenuItem_Click 函数**

openRasterLayerStripMenuItem_Click 函数实现加载栅格数据层功能，实现方法与矢量图层相应函数类似，代码如下：

```
private voidopenRasterLayerStripMenuItem_Click(object sender,
EventArgs e)
{
    string filePath = "C:\\用户目录\\栅格数据";
    string fileName = "影像-A3.tif";

    //1:创建 Raster 工作空间工厂
    IWorkspaceFactory workspcFac = new RasterWorkspaceFactory
();

    //2:打开工作空间
IRasterWorkspace rasterWorkspc;
```

```
        rasterWorkspc = workspcFac.OpenFromFile(filePath, 0)
                                                as IRasterWorkspace;
        //3:打开栅格数据集
        IRasterDataset rasterDatst = new RasterDatasetClass();
        rasterDatst = rasterWorkspc.OpenRasterDataset(fileName);

        //4:新建栅格数据层
        IRasterLayer rasterLay = new RasterLayerClass();
        rasterLay.CreateFromDataset(rasterDatst);

        //5:将新建层添加到 MapControl
        axMapControl1.AddLayer(rasterLay);
        axMapControl1.Refresh();
    }
```

**(4)添加文件浏览对话框**

以上功能目录和文件名被代码固定，不方便使用。为了解决这个问题，可以添加一个文件浏览对话框进行人机交互获取目录和文件名。

代码如下：

```
OpenFileDialog openDlg =new OpenFileDialog();
openDlg.Filter = "Shapefile (*.shp)|*.shp";
DialogResult dr = openDlg.ShowDialog();
if(dr == DialogResult.OK)
{
    string strFileName = openDlg.FileName;
    string shpPath = System.IO.Path.GetDirectoryName(strFileName);
    string shpFileName = System.IO.Path.GetFileName(strFileName);
    //to doing......
}
```

**(5)打开 File Geodatabase 要素图层**

File Geodatabase 数据库名在文件资源管理器中就是一个文件夹(格式为×××.gdb)，可以使用文件夹浏览器选定数据库，然后对应 File GDB 的工厂打开工作空间，使用工作空间的 get_DatasetNames()函数可以获得指定类型的 Dataset 名称对象的集合(它是一个枚举器)，遍历这个集合获得每个要素类的 Name 属性，以此逐一打开各要素类并添加到地图。

代码如下：

```
private voidopenFileGdbLayerStripMenuItem_Click(object sender, EventArgs e)
{
```

```
        FolderBrowserDialog openFolder = new FolderBrowserDialog();
        DialogResult dr = openFolder.ShowDialog();
        if (dr == DialogResult.OK)
        {
            string gdbName = openFolder.SelectedPath;

            //1:创建 File Geodatabase 工作空间工厂,然后打开工作空间:
            IWorkspaceFactory pWorkspaceFactory;
            pWorkspaceFactory = new FileGDBWorkspaceFactoryClass();
             IWorkspace pWorkspace = pWorkspaceFactory.OpenFromFile
(gdbName, 0);

            //2:获取工作空间 Dataset 名称对象集合,类型为简单要素
            IEnumDatasetName enumDatasetName =
                pWorkspace.get_DatasetNames(esriDatasetType.
esriDTFeatureClass);

            //遍历 DatasetName 集合;
            IFeatureWorkspace pFCWorkspace = pWorkspace as
IFeatureWorkspace;
            IDatasetName datasetName = enumDatasetName.Next();
            while (datasetName != null)
            {
                string fcName = datasetName.Name;

                //3:打开矢量数据
                IFeatureClass pFeatureClass;
                pFeatureClass = pFCWorkspace.OpenFeatureClass(fcName);

                //4:新建要素数据层
                IFeatureLayer pFeatureLayer = new FeatureLayerClass();
                pFeatureLayer.FeatureClass = pFeatureClass;
                pFeatureLayer.Name = fcName;
                axMapControl1.AddLayer(pFeatureLayer);

                //5:下一个
                datasetName = enumDatasetName.Next();
            }
```

```
        axMapControl1.Refresh();
    }
}
```

## 五、编译测试

①单击 F5 键编译运行程序;

②用【OpenShpFeatureLayer】菜单打开数据:… \\ Data \\ 制图数据 \\ :

◆ 地级城市驻地 . shp;

◆ 国界线 . shp;

◆ 省级行政区 . shp;

……

效果如图 2-1 所示。

③用【OpenRasterLayer】菜单打开数据:… \\ Data \\ 栅格数据 \\ 影像-A3. tif, 效果如图 7-1 所示。

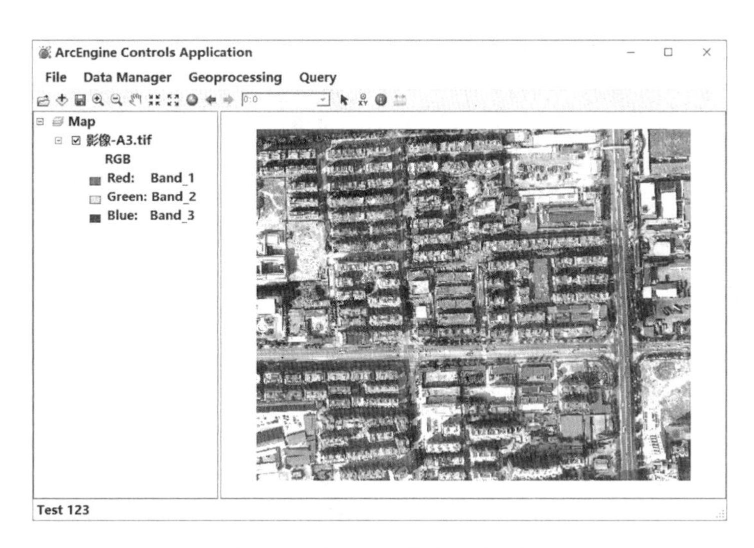


图 7-1　栅格数据加载效果图

## 六、思考与练习

①编写加载 gdb 数据库中指定要素数据集中要素类的功能。

②编写加载 mdb 数据库中要素类的功能。

# 实验八　创建要素类

## 一、目的与要求

①掌握 gdb 数据库创建方法；
②掌握使用 Workspace 创建独立要素的步骤；
③熟悉生成要素类字段定义集的要点。

## 二、实验原理

要素类(Feature Class)可以独立存储，也可以储存于要素数据集(FeatureDataset)中。独立于要素数据集的简单要素类称为独立要素类。要素数据集必须共用同一空间参考，Shapefile 文件不支持要素数据集中组织简单要素类。ArcEngine 创建要素类有两种方法：

①用工作空间 Workspace 的 CreateFeatureClass 方法创建，产生独立要素类；

②用要素数据集 FeatureDataset 的 CreateFeatureClass 方法创建，产生要素数据集的子集元素。

## 三、实验环境

①开发环境：Visual Studio 2015 + ArcGIS Engine 10.5。
②开发语言：C#。
③实验数据：... \\ Data \\ 制图数据 \\ (全国政区人口数据 shapefile)：
◆　地级城市驻地 . shp；
◆　国界线 . shp；
◆　省级行政区 . shp；
　　……

## 四、内容与步骤

本实验实现创建要素类的功能：

①用户右键单击【Data Manager】主菜单上【Create File Geodatabase】菜单项，可激活创建文件型数据库的功能。

②用户右键单击【Data Manager】主菜单上【Create Feature Class】菜单项，激活创建要素类对话框，用户可选择 gdb 数据库名、要素类名、几何类型、以及模板类等。

## 1. 创建要素类窗体设计

### (1)添加创建要素类对话框类

添加创建要素类对话框类，并命名为“CreateFeatureClassFrm”，修改窗体的 Text 属性为“Create Feature Class”，并添加 Button、Label、ComboBox 等控件。控件布局如图 8-1 所示。

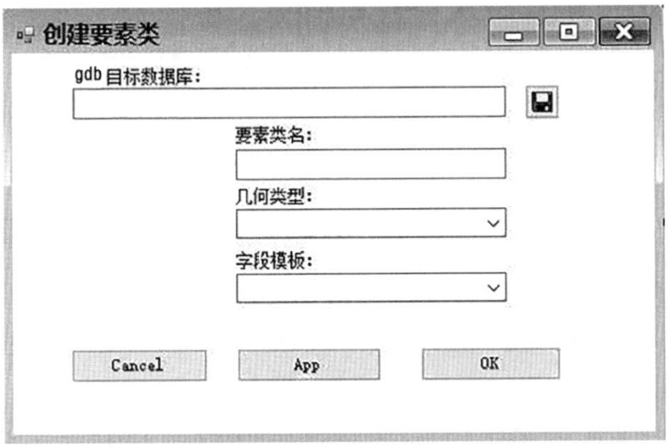

图 8-1　CreateFeatureClassFrm 控件布局图

### (2)设置控件属性

设置相应控件的相关属性，见表 8-1。

表 8-1　**CreateFeatureClassFrm 控件命名表**

| 控件 | Name 属性 | 含义 | 备注 |
|---|---|---|---|
| TextBox | txtGdbName | gdb 数据库名 | |
| TextBox | txtFCname | 新建要素类名 | |
| ComboBox | cbxShpType | 几何类型 | 集合中填写行：Point、Polyline，Polygon |
| ComboBox | cbxTemplate | 字段模板图层 | |
| Button | btnExplor | 文件浏览 | |
| Button | btnApp | 应用 | |
| Button | btnCancel | 关闭 | |
| Button | btnOK | 确定 | |

**(3)添加 CreateFeatureClassFrm 的全局变量**

```
private IMapControl3_mapControl;
```

**(4)添加 CreateFeatureClassFrm 事件的响应函数、功能函数、辅助函数**

①添加应用按钮 Click 事件响应函数;

②添加确定按钮 Click 事件响应函数。

具体代码如下:

```
public partial class CreateFeatureClassFrm : Form
{
    IMapControl3 _mapControl = null;
    public CreateFeatureClassFrm(IMapControl3 mapControl)
    {
        InitializeComponent();
        _mapControl = mapControl;
    }

    //事件响应函数
     private void CreateFeatureClassFrm _ Load ( object sender,
EventArgs e)
    private void btnExplor_Click(object sender, EventArgs e)
    private void btnOK_Click(object sender, EventArgs e)
    private void btnApp_Click(object sender, EventArgs e)
    private void btnCancel_Click(object sender, EventArgs e)

    //辅助函数
    private esriGeometryType StringToGeometryType(string
TypeOfString)
    private ILayer GetLayerByname(string lyrName)
}
```

## 2. CreateFeatureClassFrm 类的实现

**(1)Load 事件响应函数的实现**

Load 事件响应函数主要作用是:用 MapControl 中图层名填充 cbxTemplate(字段模板图层)控件,代码如下:

```
private void CreateFeatureClassFrm_Load(object sender, EventArgs
e)
{
    for (int i = 0; i < _mapControl.Map.LayerCount; i++)
    {
```

```
            ILayer aLayer = _mapControl.Map.get_Layer(i);
            if (aLayer is IFeatureLayer)
            {
                this.cbxTemplate.Items.Add(aLayer.Name);
            }
        }
    }
```

**(2) btnExplor 按钮响应函数的实现**

实现代码如下：

```
    private void btnExplor_Click(object sender, EventArgs e)
    {
        FolderBrowserDialog folderBrowserDialog1 = new
FolderBrowserDialog();
        if (folderBrowserDialog1.ShowDialog() == DialogResult.OK)
        {
            string dsName = folderBrowserDialog1.SelectedPath;
            this.txtGdbName.Text = dsName;
        }
    }
```

**(3) btnApp 按钮响应函数的实现**

btnApp 按钮响应函数完成创建要素类操作，步骤如下：

第一步，从界面获取分析参数，包括数据库名、要素类名、几何类型、字段模板等；

第二步，调用 GeodatabaseOper 功能类创建要素类，这里专门设计 GeodatabaseOper 功能类负责创建要素类工作，使程序流程更加清晰；

第三步，新要素类添加到 Map。

代码如下：

```
    private void btnApp_Click(object sender, EventArgs e)
    {
        if ( this.cbxShpType.SelectedIndex < 0 || this.txtGdbName.
Text == "" )
            return;

        string gdbPathName = System.IO.Path.GetDirectoryName(this.
txtGdbName.Text);
        string gdbFileName = System.IO.Path.GetFileName(this.
txtGdbName.Text);
        string strFCname = this.txtFCname.Text;
        string gdbTypeString = this.cbxShpType.SelectedItem.ToString
```

```
();
        esriGeometryType shapeType = StringToGeometryType(gdbType
String);

        //获取字段模板要素类
        IFeatureLayer Lyr = GetLayerByname(this.cbxTemplate.Text) as
IFeatureLayer ;
        IFeatureClass templateClass = (Lyr ! = null) ? Lyr.Feature
Class : null;

        //创建新要素类
        GeodatabaseOper fcOper = new GeodatabaseOper();
        IFeatureWorkspace fcWorkspace = fcOper.OpenFileGDbWorkspace
(gdbPathName, gdbFileName) as IFeatureWorkspace;
        IFeatureClass pFC = fcOper.CreateFeatureClass(templateClass,
fcWorkspace, strFCname, shapeType);

        //新要素类添加到 Map
        IFeatureLayer pFLayer = new FeatureLayerClass();
        pFLayer.FeatureClass = pFC;
        pFLayer.Name = strFCname;
        _mapControl.AddLayer(pFLayer);
    }

    private void btnCancel_Click(object sender, EventArgs e)
    {
        this.Close();
    }
    private void btnOK_Click(object sender, EventArgs e)
    {
        this.Close();
    }
```

**(4)辅助函数**

GetLayerByname()函数根据图名获得图层接口；StringToGeometryType()函数负责将几何类型字符串转换为几何类型枚举类型。

代码如下：

```
private ILayer GetLayerByname(string lyrName)
{
```

```
    ILayer pLayer = null;
    for (int i = 0; i < _mapControl.LayerCount; i++)
    {
        ILayer tempLayer = _mapControl.get_Layer(i);
        if (tempLayer.Name == lyrName)
        {
            pLayer = tempLayer;
            break;
        }
    }

    return pLayer;
}

private esriGeometryType StringToGeometryType(string TypeOfString)
{
    switch (TypeOfString)
    {
        case "Point":
            return esriGeometryType.esriGeometryPoint;
            break;
        case "Polyline":
            return esriGeometryType.esriGeometryPolyline;
            break;
        case "Polygon":
            return esriGeometryType.esriGeometryPolygon;
            break;
        case "Multipoint":
            return esriGeometryType.esriGeometryMultipoint;
            break;
        case "MultiPatch":
            return esriGeometryType.esriGeometryMultiPatch;
            break;
        default:
            return esriGeometryType.esriGeometryPoint;
            break;
    }
}
```

### 3. 功能类 GeodatabaseOper 实现

GeodatabaseOper 类设计提供 CreateFeatureClass( )供外部调用，涉及创建字段辅助函数和创建空间参考辅助函数：

①CreateDefaultFields( )函数：创建默认字段定义集；

②CloneFeatureClassFields( )函数：克隆一个字段定义集；

③CreateDefalutSpatialRef( )函数：创建一个默认空间参考；

④CloneSpatialRef( )函数：克隆一个空间参考。

代码如下：

```
class GeodatabaseOper
{
    //公有函数
    public IFeatureClass CreateFeatureClass(object template,
    IFeatureWorkspace fcWorkspace, string strFCname,
     esriGeometryType shapeType)

    //字段创建函数
    public IFields CreateDefaultFields(esriGeometryType shapeType,
          ISpatialReference sr)
    public IFields CloneFeatureClassFields(IFeatureClass pfc,
          esriGeometryType geoType, ISpatialReference sr)

    //空间参考创建函数
    public ISpatialReference CloneSpatialRef(IFeatureClass templateFC)
      public ISpatialReference CreateDefalutSpatialRef( )

    //辅助函数
    private IGeometryDef CreateGeometryDef(esriGeometryType shapeType,
      ISpatialReference spatialReference2)

}
```

**(1) CreateFeatureClass( )函数**

CreateFeatureClass( )函数实现的基本思路是：

第一步，创建字段定义集合(这是创建要素类的难点所在)，如果模板是一个要素类，则克隆一个字段定义集合，用到 CloneFeatureClassFields( )函数，调用之前需要克隆空间参考系(CloneSpatialRef( ))。如果模板是字段集或者为空，则使用 CreateDefaultFields( )函数创建新的字段定义集合，并在此之前要创建一个默认参考系(CreateDefaultSpatialRef( )，坐标系是 Unknown)；

第二步，调用工作空间的 CreateFeatureClass( )函数完成要素类创建。

代码如下：

```
public IFeatureClass CreateFeatureClass(object template,
    IFeatureWorkspace fcWorkspace,string strFCname, esriGeometry
Type shapeType)
{
  //创建字段定义集合
    IFields pFields = null;
    if (template is IFeatureClass )
    {
        //克隆空间参考
        ISpatialReference targetSR = CloneSpatialRef(template as
IFeatureClass); ;
        pFields = CloneFeatureClassFields(template as IFeature
Class, shapeType, targetSR);
    }
    else if (template is IFields)
    {
        //创建默认参考系
        ISpatialReference targetSR = CreateDefalutSpatialRef( );
        pFields = CreateDefaultFields(template as IFields, shape
Type, targetSR);
    }
    else
    {
        //创建默认参考系
        ISpatialReference targetSR = CreateDefalutSpatialRef();
        pFields = CreateDefaultFields(null, shapeType, targetSR);
    }

    //创建要素类
     return fcWorkspace.CreateFeatureClass ( strFCname, pFields,
null, null,esriFeatureType.esriFTSimple, "Shape", "");
}
```

**(2) CreateDefaultFields( )，CreateGeometryDef 函数**

字段定义集合里，必须单独定义两个特殊的字段，一是对象标识字段(esriFieldTypeOID)，二是几何图形字段(esriFieldTypeGeometry)，其他字段为“用户自定义字段”，可通过 userFields 参数给定。如果用户没有给定，则创建一个字符串型的【NAME】字段。

对于几何图形字段，必须为字段的 IGeometryDef 属性赋值，该属性可确定几何图形的几何类型（Point、Polyline、Polygon 等），空间参考系，空间索引等，这里通过 CreateGeometryDef()函数创建 IGeometryDef 属性，为简化起见，空间索引使用固定数值。

代码如下：

```
public IFields CreateDefaultFields (IFields userFields,
                esriGeometryType shapeType,ISpatialReference sr)
{
    //创建新的字段集
    IFields pFields = new FieldsClass();
    IFieldsEdit pFieldsEdit = (IFieldsEdit)pFields;
    //产生新的 FID 字段
    {
        IField pField = new FieldClass();
        IFieldEdit pFieldEdit = (IFieldEdit)pField;
        pFieldEdit.Name_2 = "FID";
        pFieldEdit.AliasName_2 = "FID";
        pFieldEdit.Type_2 = esriFieldType.esriFieldTypeOID;
        pFieldsEdit.AddField(pField);
    }
    //产生新的 shape 字段
    {
        IField pField = new FieldClass();
        IFieldEdit pFieldEdit = (IFieldEdit)pField;
        pFieldEdit.Name_2 = "Shape";
        pFieldEdit.AliasName_2 = "Shape";
        pFieldEdit.Type_2 = esriFieldType.esriFieldTypeGeometry;
        pFieldEdit.GeometryDef_2 = CreateGeometryDef(shapeType,
sr);
        pFieldsEdit.AddField(pField);
    }

    //产生自定义字段:根据需要确定
    if (userFields ! = null)
    {
        for (int i = 0; i < userFields.FieldCount; i++)
```

```
        {
            IField pField = userFields.get_Field(i);
            pFieldsEdit.AddField(pField);
        }
    }
    else
    {
        IField pField = new FieldClass();
        IFieldEdit pFieldEdit = (IFieldEdit)pField;
        pFieldEdit.Name_2 = "NAME";
        pFieldEdit.AliasName_2 = "名称";
        pFieldEdit.Type_2 = esriFieldType.esriFieldTypeString;
        pFieldsEdit.AddField(pField);
    }

    return pFields;
}

private IGeometryDef CreateGeometryDef(esriGeometryType shapeType,
ISpatialReference spatialReference2)
{
    IGeometryDef geometryDef = new GeometryDefClass();
    IGeometryDefEdit geometryDefedit = (IGeometryDefEdit)
geometryDef;
    {
        //空间数据类型
        geometryDefedit.GeometryType_2 = shapeType;
        //空间参考系
        geometryDefedit.SpatialReference_2 = spatialReference2;
        //平均点数
        geometryDefedit.AvgNumPoints_2 = 5;
        //空间索引级数
        geometryDefedit.GridCount_2 = 1;
        //第一级索引格大小
        geometryDefedit.set_GridSize(0, 10000);
    }
```

```
    return geometryDef;
}
```

**(3) CloneFeatureClassFields( )函数**

第一步，先通过 IFieldChecker 接口获得模板要素类的有效字段集(validateFields)；

第二步，遍历有效字段集，如果字段不是 Shape 类型，则添加到新建字段集(pFields)，否则修改该字段的 IGeometryDef 属性的几何类型和空间参考，然后再添加到新建字段集。

代码如下：

```
private IFields CloneFeatureClassFields(IFeatureClass pfc, esri
            GeometryType geoType,ISpatial Reference sr)
{
    //获得有效字段集
    IFieldChecker fieldChecker = new FieldCheckerClass();
    IEnumFieldError enumFieldError = null;
    IFields validateFields = null;
    fieldChecker.ValidateWorkspace = (pfc as IDataset).Workspace;
    fieldChecker.Validate(pfc.Fields, out enumFieldError, out
validateFields);

    //构造新的字段集
    IFields pFields = new FieldsClass();
    IFieldsEdit pFieldsEdit = (IFieldsEdit)pFields;

    //根据传入的要素类有效字段集,复制除了 shape 字段之外的字段
    long inFieldsCount = validateFields.FieldCount;
    long indexShape = validateFields.FindField(pfc.Shape
FieldName);
    for (int i = 0; i < inFieldsCount; i++)
    {
        IField pField = validateFields.get_Field(i);

        //修改 IField 字段的 GeometryDef 属性
        if (i == indexShape)
        {
            IFieldEdit pFieldEdit = (IFieldEdit)pField;

            //克隆一个 IGeometryDef;
            IClone defClone = (pFieldEdit.GeometryDef as IClone)
```

```
.Clone();
                IGeometryDef geometryDef = defClone as IGeometryDef;

                //修改 IGeometryDef 的 GeometryType/SpatialReference 属性
                IGeometryDefEdit geometryDefedit = (IGeometryDefEdit)
geometryDef;
                geometryDefedit.GeometryType_2 = geoType;
                geometryDefedit.SpatialReference_2 = sr;

                //修改 IField 的 IGeometryDef 属性
                pFieldEdit.GeometryDef_2 = geometryDef;

            }

            pFieldsEdit.AddField(pField);
        }

        return pFields;
    }
```

(4) **CreateDefaultSpatialRef()、CloneSpatialRef()函数实现**

实现代码如下：

```
    public ISpatialReference CloneSpatialRef(IFeatureClass
templateFC)
    {
        ISpatialReference targetSR = null;
        if (templateFC ! = null)
        {
            //克隆空间参考
            ISpatialReference spatialRefer = (templateFC as
IGeoDataset).SpatialReference;
            IClone srClone = (spatialRefer as IClone).Clone();
            targetSR = srClone as ISpatialReference;
        }

        return targetSR;
    }

    public ISpatialReference CreateDefalutSpatialRef( )
```

```
    {
        //创建一个 Unknown 坐标系
        ISpatialReference targetSR = null;
        targetSR = new UnknownCoordinateSystemClass();
        targetSR.SetDomain(-100000000, 100000000, -100000000,
100000000);

        //设置容差
        ISpatialReferenceTolerance srTolerance = targetSR as
ISpatialReferenceTolerance;
        srTolerance.XYTolerance = 0.001;
        srTolerance.ZTolerance = 0.001;

        //设置分辨率
        ISpatialReferenceResolution srResolution = targetSR as
ISpatialReferenceResolution;
        srResolution.ConstructFromHorizon();
        srResolution.SetDefaultXYResolution();

        return targetSR;
    }
```

**(5) gdb 数据库创建/打开函数**

为方便使用，GeodatabaseOper 类提供 gdb 数据库的创建、打开函数。

代码如下：

```
    public IWorkspace CreateFileGDbWorkspace (string path, string
gdbName)
    {
        IWorkspace workspace = null;
        try
        {
      Type factoryType =
    Type.GetTypeFromProgID("esriDataSourcesGDB.FileGDBWorkspace
Factory");
      IWorkspaceFactory workspaceFactory =
    (IWorkspaceFactory)Activator.CreateInstance(factoryType);

           //创建工作名字空间
     IWorkspaceName workspaceName = workspaceFactory.Create(path,
```

```
gdbName, null, 0);

      IName name = (IName)workspaceName;
      workspace = (IWorkspace)name.Open();
        }
        catch (Exception Err)
        {
            MessageBox.Show(Err.ToString());
            workspace = null;
        }

        return workspace;
    }

    public IWorkspace OpenFileGDbWorkspace (string path, string
gdbName)
    {
        string strFullname = path + "\\" + gdbName;
        IWorkspace workspace = null;
        IWorkspaceFactory workspaceFactory = null;
        if (Directory.Exists(strFullname))
        {
            //打开工作空间
            //Instantiate an Access workspace factory and create a
personal geodatabase.
            //The Create method returns a workspace name object.
            try
            {
                Type factoryType =
    Type.GetTypeFromProgID("esriDataSourcesGDB.FileGDBWorkspace
Factory");
                workspaceFactory = (IWorkspaceFactory)Activator.
CreateInstance(factoryType);
                //open workspace
                workspace=workspaceFactory.OpenFromFile(strFullname,0);
            }
            catch (Exception Err)
            {
```

```
                MessageBox.Show(Err.ToString());
                workspace = null;
            }
        }

        return workspace;
    }

    public IWorkspace OpenShapefileWorkspace(string path)
    {
        string strFullname = path ;
        IWorkspace workspace = null;
        IWorkspaceFactory workspaceFactory = null;
        if (Directory.Exists(strFullname))
        {
            //打开工作空间
            //Instantiate an Access workspace factory and create a
personal geodatabase.
            //The Create method returns a workspace name object.
            try
            {
                Type factoryType =
    Type.GetTypeFromProgID("esriDataSourcesFile.ShapefileWorkspace
Factory");
                workspaceFactory = (IWorkspaceFactory)Activator.
CreateInstance(factoryType);
                //open workspace
                workspace = workspaceFactory.OpenFromFile(strFullname,
0);
            }
            catch (Exception Err)
            {
                MessageBox.Show(Err.ToString());
                workspace = null;
            }
        }

        //循环释放直至返回0
```

```
        //WhileReleaseWorkspaceFactory(workspaceFactory);
        return workspace;
    }
```

### 4. 功能调用

①在主菜单【Data Manager】上添加一菜单项(命名为 Create File Geodatabase)，创建并修改 Click 事件响应函数，代码如下：

```
    private void createFileGeodatabaseToolStripMenuItem_Click
(object sender, EventArgs e)
    {
        SaveFileDialog saveFileDialog1 = new SaveFileDialog();
        saveFileDialog1.Filter = "gdbfile (*.gdb)|*.gdb";
        if (saveFileDialog1.ShowDialog() == DialogResult.OK)
        {
            string txtGdbName = saveFileDialog1.FileName;
            string gdbPathName = System.IO.Path.GetDirectoryName
(txtGdbName);
            string gdbFileName = System.IO.Path.GetFileName
(txtGdbName);
            try
            {
                GeodatabaseOper gdbOper = new GeodatabaseOper();
                gdbOper.CreateFileGDbWorkspace(gdbPathName,
gdbFileName);
                MessageBox.Show("Create Gdb is Finished");
            }
            catch(Exception ex)
            {
                MessageBox.Show("Create Gdb is Failed");
            }
        }
    }
```

②在主菜单【Data Manager】上添加一菜单项(命名为 Create Feature Class)，创建并修改 Click 事件响应函数，代码如下：

```
    private void createFeatureClassToolStripMenuItem1_Click(object
sender, EventArgs e)
    {
        CreateFeatureClassFrm frm = new CreateFeatureClassFrm(m_
```

```
mapControl);
            if (frm.ShowDialog() = = DialogResult.OK)
            {
                this.m_mapControl.ActiveView.Refresh();
            }
      }
```

## 五、功能测试

①点击菜单【Create File Geodatabase】启动 gdb 数据库创建对话框，在目录浏览里选择“存储目录名”、“数据库名”，点击【OK】按钮提示“创建成功”或“失败”。

②点击菜单【Create Feature Class】启动要素类创建对话框；

- 设置 gdb 数据库(在 Windows 资源管理器中就是目录)，选择几何类型等参数；
- 点击【应用】按钮，确定后即可看到新建的要素类已添加到地图。

## 六、思考与练习

①试在 gdb 数据库中创建要素数据集的功能。

②试在要素数据集上创建要素类。

# 实验九　几何对象基本操作

## 一、目的与要求

①熟悉 ArcGIS Engine Geometry 对象模型常用接口的用法；
②熟悉 ArcGIS Engine IFeatureClass 相关接口的用法；
③熟悉 Windows 文件流读取文本数据文件的方法。

## 二、实验原理

了解 ArcGIS 几何对象模型需要从 Geometry 对象模型图入手，如图 9-1 所示。

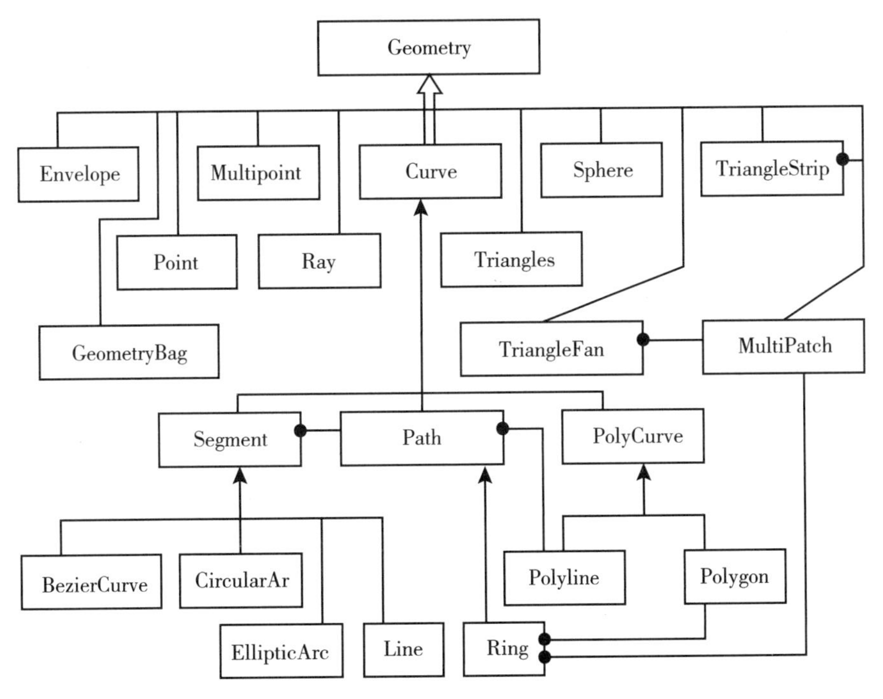


图 9-1　Geometry 对象模型图

①Segment 对象是一个有起点和终点的“线段”，也就是说 Segment 只有两个点，至于

两点之间的线是直的，还是弯曲的，需要其他的参数定义。所以，Segment 是由起点、终点和参数三个方面决定的。Segment 有 4 个子类，它的 4 个子类分别是直线、圆弧、椭圆弧、贝赛尔曲线。

②Path 是连续(首尾相接)的 Segment 的集合，除了路径的第一个 Segment 和最后一个 Segment 外，其余的 Segment 的起始点都是前一个 Segment 的终止点，当 Path 封闭时(即起始和终止点有相同的坐标值)就是 Ring。

Path/Ring 实现了 IPointCollection 接口提供直接对节点操作的功能；对于简单的 Path 对象(直线段组成)，可用这个接口直接创建 Path。

③Polyline/Polygon 是 Path/Ring 组成的有序集合，IGeometryCollection 接口提供了访问 Polyline/Polygon 中的 Path/Ring 的方法。

一个 Polyline 对象必须满足以下准则：

◆ 组成 Polyline 的 Path 对象都是有效的；
◆ Path 不会重合、相交或自相交；
◆ 多个 Path 对象可以连接于某一个节点，也可以是分离的；
◆ 长度为 0 的 Path 对象是不被允许的。

一个 Polygon 对象必须满足以下准则：

◆ 组成 Polygon 对象的每一个 Ring 都是有效的；
◆ Ring 之间的边界不能重合；
◆ 外部环是有方向的，它是顺时针方向；
◆ 内部环在一个多边形中定义了一个洞，它是逆时针方向的；
◆ 面积为 0 的环是不允许的。

④Geometry 对象分为两个层次，一类是直接用来构成要素的 Geometry 对象，二类是构成这些对象的部件对象，前者称为高级几何对象(Polyline、Polygon、Point 等)，后者称为低级几何对象(Path、Ring、Segment 等)。

⑤为保证 Polyline/Polygon 的合法性，在构造时可调用 ITopologicOperator 接口的 Simplify()函数进行合法化处理。注意：只有高级几何对象才能实现这个接口，如果需要对低级几何对象进行合法化操作，可将其包装成高级几何对象。

## 三、实验环境

①开发环境：Visual Studio 2005；

②开发语言：C#；

③实验数据：… \\ Data \\ COR 数据 \\ ：

◆ 实验_FZUBGIS_500-A. Cor；
◆ 实验_FZUBGIS_500-L. Cor；
◆ 实验_FZUBGIS_500-P. Cor。

# 四、内容与步骤

本实验通过读取 EPSCOR 数据文件，实现构建 Polygon 对象的基本功能。

EPSCOR 文件是 EPS 数字测图软件支持的一种较简单的 GIS 数据文本格式，按分块记录几何图形实体，以空格作为分隔符，每块格式如下：

【编码】【点数】【线型】　　//块头

【点 id】【E】【N】【H】　　//数据体

…（多行，每个点一行）

预期功能是：用户点击主菜单【Data Manager】中【Convert Cor File】菜单项，出现 Cor 文件转换对话框，选择 COR 格式数据文件，设置输出 shp 文件等，点击确认后，程序将读取 Cor 文件中多边形数据块，构建成 Polygon 要素添加到指定 shp 文件中。

## 1. 建立窗体

### (1) 新建窗体

在 Windows 中新建一个窗体，并命名为"CortoShapefileFrm"，窗体控件布局如图 9-2 所示。

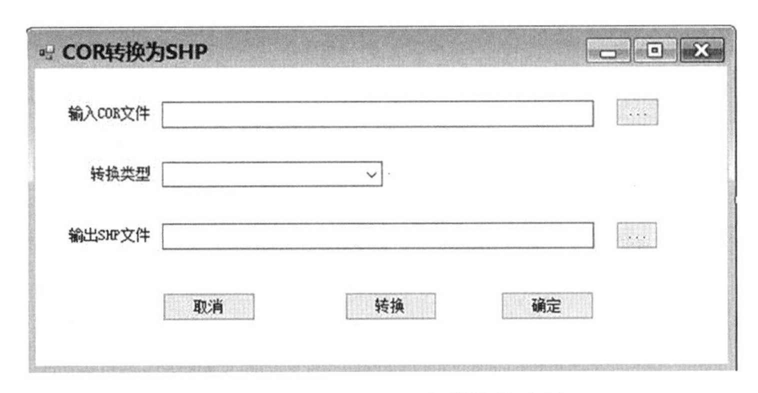


图 9-2　CortoShapefile 窗体控件布局

### (2) 设置控件属性

设置相应控件的相关属性，见表 9-1。

表 9-1　**CortoShapefile 窗体控件命名表**

| 控件 | Name 属性 | 含义 | 备注 |
|---|---|---|---|
| TextBox | txtCorName | Cor 文件名 | |
| TextBox | txtOutputSHP | 输出 shp 文件名 | |
| ComboBox | cbxshpType | 几何类型 | 集合中填写行：Point、Polyline、Polygon |

续表

| 控件 | Name 属性 | 含义 | 备注 |
| --- | --- | --- | --- |
| Button | btnExplor1 | 输入文件浏览 | |
| Button | btnExplor2 | 输出文件浏览 | |
| Button | btnApply | 转换 | |
| Button | btnCancel | 关闭 | |
| Button | btnOK | 确定 | |

**(3)类设计**

代码如下：

```
public partial class CortoShapefileFrm : Form
{
    object obj = Type.Missing;
    private IMapControl3 m_mapControl = null;
    public CortoShapefileFrm(IMapControl3 mapControl)
    {
        InitializeComponent();
        m_mapControl = mapControl;
    }

    //输入文件浏览响应函数
    private void btnExplor1_Click(object sender, EventArgs e)
    //输出文件浏览响应函数
    private void btnExplor2_Click(object sender, EventArgs e)
    //输入文件变化响应函数
    private void txtCorName_TextChanged(object sender, EventArgs e)
    //执行响应函数
    private void btnApply_Click(object sender, EventArgs e)
    private void btnCancel_Click(object sender, EventArgs e)
    private void btnOK_Click(object sender, EventArgs e)

    //功能函数:将 COR 文件的面图层转为目标要素类
     private void SaveTo_TargetFeatueClas_A(StreamReader sr,
IFeatureClass targetFC)

    //辅助函数:创建一个 shp 要素类
      private  IFeatureClass  CreateFeatureClassForCor ( string
```

```
shpFullName,esriGeometryType geoType)
        //辅助函数:读取一个环
        private IRing ReadRingByCount(StreamReader sr, int
nPointCount)
        //辅助函数:读取文件中一行字符串
        private bool ReadLineString(StreamReader sr, ref string str)
        //辅助函数:将几何类型字符串转换为几何类型枚举类型
        private esriGeometryType StringToGeometryType(string
TypeOfString)
    }
```

**(4)响应函数实现**

btnApply_Click()函数负责调度转换流程，步骤如下：

①创建目标要素：用到辅助函数 CreateFeatureClassForCor()；

②获取一个指向文件流的流读取器；

③根据目标类型，存储 COR 数据到目标要素类中，本实验仅实现 polygon 类型转换，其他类型请读者自行完成；

④添加目标要素类到图层。

具体代码如下：

```
    private void btnApply_Click(object sender, EventArgs e)
    {
        //创建目标要素
        esriGeometryType geometryType = StringToGeometryType(this.
cbxshpType.Text);
        IFeatureClass targetFC =CreateFeatureClassForCor(txtOutputSHP.
Text, geometryType );

        //获取一个指向文件流的流读取器
        FileInfo fi = new FileInfo(txtCorName.Text);
        FileStream fs = fi.Open(FileMode.Open, FileAccess.Read,
FileShare.ReadWrite);
        StreamReader sReader = new StreamReader(fs, Encoding.
GetEncoding("gb2312"));

        //存储 COR 数据到目标要素中
        switch (geometryType)
        {
            case esriGeometryType.esriGeometryPolygon:
                SaveTo_TargetFeatueClas_A(sReader, targetFC);
```

```
            break;
        case esriGeometryType.esriGeometryPolyline:
            //读者自行完成
            //SaveToTargetFeatueClas_L(sReader, targetFC);
            break;
        case esriGeometryType.esriGeometryPoint:
        default:
            //读者自行完成
            //SaveToTargetFeatueClas_P(sReader, targetFC);
            break;
            }

    //关闭文件流
    sReader.Close();
    fs.Close();

    //添加目标要素类到图层
    IFeatureLayer pfeatureLyr = new FeatureLayerClass();
    pfeatureLyr.FeatureClass = targetFC;
    pfeatureLyr.Name = targetFC.AliasName;
    m_mapControl.AddLayer(pfeatureLyr, 0);
}
```

其他响应函数，代码如下：

```
private void btnExplor1_Click(object sender, EventArgs e)
{
    //打开工作空间
    OpenFileDialog openDlg = new OpenFileDialog();
    openDlg.Filter = "COR file ( * .cor) |* .cor";
    DialogResult dr = openDlg.ShowDialog();
    if (dr == DialogResult.OK)
    {
        string strFileName = openDlg.FileName;
        txtCorName.Text = strFileName;
    }
}

//点击输出路径按钮时,执行函数
```

```
private void btnExplor2_Click(object sender, EventArgs e)
{
    //set the output layer
    SaveFileDialog saveDlg = new SaveFileDialog();
    saveDlg.CheckPathExists = true;
    saveDlg.Filter = "Shapefile ( * .shp) |* .shp";
    saveDlg.OverwritePrompt = true;
    saveDlg.Title = "Output Layer";
    saveDlg.RestoreDirectory = true;
    saveDlg.FileName = " ";
    DialogResult dr = saveDlg.ShowDialog();
    if (dr == DialogResult.OK)
        txtOutputSHP.Text = saveDlg.FileName;
}

private void txtCorName_TextChanged(object sender, EventArgs e)
{
    int index = txtCorName.Text.IndexOf(".");
    string strName = txtCorName.Text.Substring(0, index);
    txtOutputSHP.Text = strName + ".shp";
}

private void btnCancel_Click(object sender, EventArgs e)
{
    this.Close();
}

private void btnOK_Click(object sender, EventArgs e)
{
    this.Close();
}
```

**(5)功能函数 SaveTo_TargetFeatueClas_A()的实现**

SaveTo_TargetFeatueClas_A()用辅助函数 ReadLineString()读取 COR 文件中的文本，每取得一个块头，用辅助函数 ReadRingByCount()读取一个 Ring 实体，然后将 Ring 包装成一个 Polygon 创建一个新要素，并将从块头分解到的编码信息存储到新要素的【CODE】字段中。

本例仅支持 Polygon 类型 COR 文件转换到 SHP 文件。读者可参考此例，编写支持

Polyline 和 Point 类型的函数。

代码如下：

```
private void SaveTo_TargetFeatueClas_A(StreamReader sr, IFeature
Class targetFC)
{
    int iCode = targetFC.Fields.FindField("CODE");
    string str = "";
    object obj = Type.Missing;

    //将每个 COR 数据块转换为一个要素
    while (ReadLineString(sr, ref str))
    {
        string[] strArr = str.Split(new char[] {' '});
        int nCode = int.Parse(strArr[0]);
        int nPointCount = int.Parse(strArr[1]);
        IRing ring = ReadRingByCount(sr, nPointCount);
        if (ring ! = null)
        {
            IGeometryCollection polygon = new PolygonClass();
            polygon.AddGeometry(ring, ref obj, ref obj);

            //创建新要素
            IFeature pFeature = targetFC.CreateFeature();
            pFeature.Shape = polygon as IGeometry; //保存图形
            pFeature.set_Value(iCode, nCode);      //保存编码信息
            pFeature.Store();
        }
        else
        {
            break;
        }
    }
}
```

**(6)辅助函数**

CreateFeatureClassForCor()函数创建一个与 COR 文件内容匹配的要素类(包含自定义字段：【CODE】)，这里用到 GeodatabaseOper 类，参见实验《创建要素类》。

```
private IFeatureClass CreateFeatureClassForCor(string shpFullName,
```

```
                                        esriGeometryType geoType)
    {
        //获取 shp 工作空间的目录名、shapefile 文件名,打开工作空间
        string shpPathName = System.IO.Path.GetDirectoryName
(shpFullName);
        string shpFileName = System.IO.Path.GetFileName
(shpFullName);
        GeodatabaseOper fcOper = new GeodatabaseOper();
        IFeatureWorkspace fcWorkspace = fcOper.OpenShapefile
Workspace(shpPathName) as IFeatureWorkspace;

        //创建"CODE"字段描述
        IFields pFields = new FieldsClass();
        IFieldsEdit pFieldsEdit = (IFieldsEdit)pFields;
        {
            IField pField = new FieldClass();
            IFieldEdit pFieldEdit = (IFieldEdit)pField;
            pFieldEdit.Name_2 = "CODE";
            pFieldEdit.AliasName_2 = "CODE";
            pFieldEdit.Type_2 = esriFieldType.esriFieldTypeInteger;
            pFieldsEdit.AddField(pField);
        }

        //创建新要素类
        return fcOper.CreateFeatureClass(pFields, fcWorkspace,
shpFileName, geoType);
    }
```

ReadRingByCount()函数读取一个数据块，返回一个环，代码如下:

```
    //读取一个环
    private IRing ReadRingByCount(StreamReader sr, int nPointCount)
    {
        IPointCollection pointList = new RingClass();
        for (int i = 0; i < nPointCount; i++)
        {
            string str = "";
            if (! ReadLineString(sr, ref str))
```

```
            return null;
        string[] strArr = str.Split(new char[] {' '});
        if (strArr.Length < 3)
            return null;
        double x = double.Parse(strArr[1]);
        double y = double.Parse(strArr[2]);
        //double z = double.Parse(strArr[3]);
        IPoint point = new PointClass();
        point.PutCoords(x, y);
        object obj = Type.Missing;
        pointList.AddPoint(point, ref obj, ref obj);
    }
    IRing ring = pointList as IRing;
    ring.Close();
    return ring;
}

//读取一行文本
private bool ReadLineString(StreamReader sr, ref string str)
{
    while (! sr.EndOfStream)
    {
        str = sr.ReadLine();
        str.Trim();
        if (str.Length == 0)
            continue;
        else
            return true;
    }
    return false;
}

//StringToGeometryType()将字符串转换为 GeometryType 枚举类型
private esriGeometryType StringToGeometryType(string TypeOfString)
{
    switch (TypeOfString)
    {
```

```
        case "Point":
            return esriGeometryType.esriGeometryPoint;
            break;
        case "Polyline":
            return esriGeometryType.esriGeometryPolyline;
            break;
        case "Polygon":
            return esriGeometryType.esriGeometryPolygon;
            break;;
        default:
            return esriGeometryType.esriGeometryPoint;
            break;
    }
}
```

2. 功能调用

在【Data Manager】主菜单上添加【Convert Cor File】菜单项，并创建响应函数，代码如下：

```
private voidconvertCorFileToolStripMenuItem_Click(object sender,
EventArgs e)
{
    CortoShapefileFrm frm =new CortoShapefileFrm(m_mapControl);
    frm.ShowDialog();
}
```

## 五、编译测试

①单击 F5 键，编译运行程序；

②点击菜单【Convert Cor File】启动 COR 转换对话框，选择：

- ◆ COR 文件名= ... \\ Data \\ COR 数据 \\ 实验_FZUBGIS_500-A. Cor;
- ◆ 指定几何类型= Polygon;

③点击【应用】，生成实验_FZUBGIS_500-A. shp 文件，并加载到地图。

效果如图 9-3 所示。

## 六、思考与练习

①如何读取点数据文件到 Shapefile 文件?

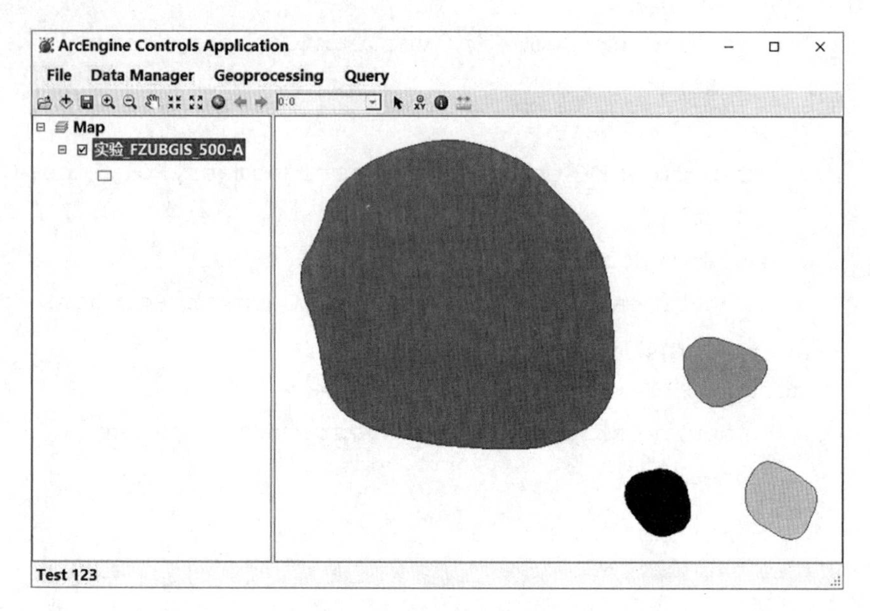


图 9-3　COR 转换效果图

②如何读取线要素文件到 Shapefile 文件?

# 实验十　空间数据属性表显示

## 一、目的与要求

①熟悉使用要素游标(ICursor 接口)遍历要素的基本方法；
②熟悉要素字段定义集(IFields)组成，以及如何获取指定字段的值；
③熟悉用 DataGridView 展示要素图层的属性表。

## 二、实验原理

从关系数据库的角度而言，ITable 对象代表了 GeoDatabase 中的一张二维表或者视图 View，一个表由多个列(IFields)定义，列被称为字段(IField)，IFields 是 IField 集合。使用 ITable 对象添加字段(AddField)、删除字段(DeleteField)等方法，可改变表的结构。存储在表中的元素是 IRow 对象，一个 IRow 对象代表了表中的一条记录。

IFeatureClass 对象继承了 ITable，所以它也是一个表，但它只有一个表示要素空间图形的特殊字段(Geometry)，一行记录代表一个要素。对于离散型栅格数据(IRaster)，每个波段存在(或可重建)表示"值-频数"统计信息的属性表(AttributeTable)，它属于 ITable 类型，因此我们也可以用表格的形式打开一个要素类或打开栅格数据的属性表。

DataGridView 控件提供了一种强大而灵活的以表格形式显示数据的方式。可以使用 DataGridView 控件来显示含少量数据的只读视图，也可以对其进行缩放以显示特大数据集的可编辑视图，还可以很方便地把一个 DataTable 绑定到 DataGridView 控件数据源。利用 DataGridView 显示 ITable 数据，比较有效的办法是：先将 ITable 转换为 DataTable，包括创建 DataTable 结构，填充表体等；然后再将 DataTable 绑定到 DataGridView 控件数据源；最后调用并显示属性表窗体。

## 三、实验环境

①开发环境：Visual Studio 2015 + ArcGIS Engine 10.5；
②开发语言：C#；
③实验数据：.. \\ Data \\ 制图数据 \\ (全国政区人口数据 shapefile)：
- ◆ 地级城市驻地 . shp；
- ◆ 国界线 . shp；

◆ 省级行政区 . shp。

## 四、内容与步骤

在 ArcMap 中，单击图层浮动菜单中的“Open Attribute Table”命令，便可弹出属性数据表。本实验将完成类似的功能，效果如图 10-1 所示。

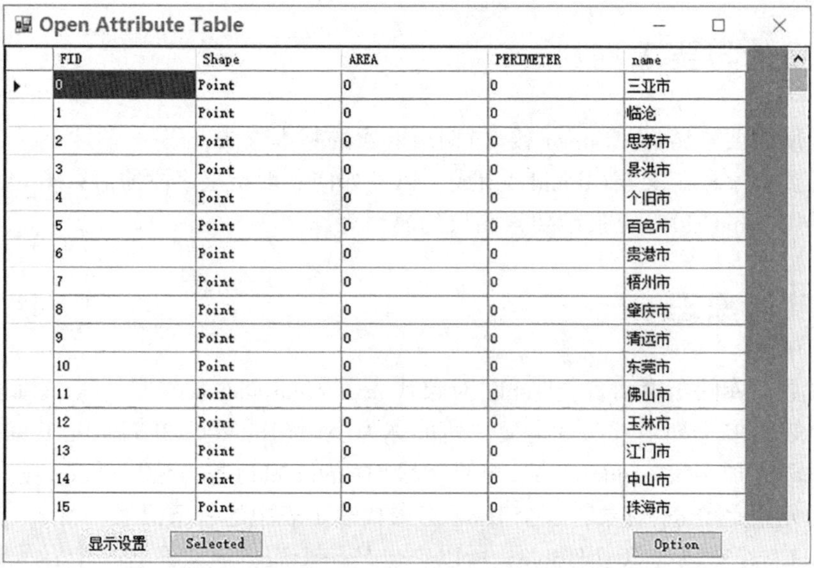

Open Attribute Table

| FID | Shape | AREA | PERIMETER | name |
|---|---|---|---|---|
| 0 | Point | 0 | 0 | 三亚市 |
| 1 | Point | 0 | 0 | 临沧 |
| 2 | Point | 0 | 0 | 思茅市 |
| 3 | Point | 0 | 0 | 景洪市 |
| 4 | Point | 0 | 0 | 个旧市 |
| 5 | Point | 0 | 0 | 百色市 |
| 6 | Point | 0 | 0 | 贵港市 |
| 7 | Point | 0 | 0 | 梧州市 |
| 8 | Point | 0 | 0 | 肇庆市 |
| 9 | Point | 0 | 0 | 清远市 |
| 10 | Point | 0 | 0 | 东莞市 |
| 11 | Point | 0 | 0 | 佛山市 |
| 12 | Point | 0 | 0 | 玉林市 |
| 13 | Point | 0 | 0 | 江门市 |
| 14 | Point | 0 | 0 | 中山市 |
| 15 | Point | 0 | 0 | 珠海市 |

显示设置 Selected Option

图 10-1 属性表显示效果图

**(1)创建属性表窗体**

新建一个 Windows 窗体，命名为“OpenAttributeTableFrm. cs”。

从工具箱中拖一个 DataGridView 控件到窗体，并将其 Dock 属性设置为“Fill”。

代码如下：

```
public partial class OpenAttributeTableFrm: Form
{
    privateIFeatureLayer m_pFeatureLayer = null;
    publicOpenAttributeTableFrm(IFeatureLayer pFeatureLayer)
    {
        InitializeComponent();
        m_pFeatureLayer = pFeatureLayer;
    }

    public System.Windows.Forms.DataGridView AttrGridView
    {
```

```
            get { return this.dataGridView1; }
        }

        private void OpenAttributeTableForm_Load(object sender, 
EventArgs e)
        {
            AttrGridView.DataSource = this.GenerateDataTable(m_
pFeatureLayer);
        }
    }
```

**(2)数据源生成函数 GenerateDataTable()**

①初始化一个 DataTable;

②IFeatureClass 接口包含 Fields 属性(字段定义集),遍历 Fields 每个字段(Field),用字段名称(name)建立 DataTable 表头;

③使用 ITable 的 Search 函数获得游标 ICursor;使用游标遍历要素类,用要素的属性值填充 DataTable 表的行元素,其中:游标 NextRow()函数返回 IRow 接口,IRow 的 get_Value(i)函数获得第 i 个字段的值,该值赋给行元素的第 i 列。

类似地,用 IFeatureSelection 取得 ISelectionSet 接口,通过游标 ICursor 来访问"选择集"中的要素,请学生自己完成。具体实现代码如下:

```
    //用数据层创建一个表
    public DataTable GenerateDataTable(IFeatureLayer pFealyr)
    {
        IFeatureClass pFCls = pFealyr.FeatureClass;
        IFields pFields = pFCls.Fields;

        //填充表头
        DataTable pTable = new DataTable();
        for (int i = 0; i < pFields.FieldCount; i++)
        {
            pTable.Columns.Add(pFields.get_Field(i).Name);
        }

        //通过要素游标填充表
        ICursor pCursor = (pFCls as ITable).Search(null,true);
        IRow pFea = pCursor.NextRow();
        while (pFea != null)
        {
            //初始化 DataTable 新行
```

```
            DataRow pRow = pTable.NewRow();
            //填充 DataTable 新行
            for (int i = 0; i < pFields.FieldCount; i++)
            {
                pRow[i] = pFea.get_Value(i).ToString();
            }

            //DataTable 新行添加到行集
            pTable.Rows.Add(pRow);

            //下一个
            pFea = pCursor.NextRow();
        }

        return pTable;
    }
```

**(3)调用属性表窗体**

要在 TOCControl 浮动菜单中添加菜单项：OpenAttributeTable。建立 OpenAttributeTable 的 OnClick 事件的响应代码如下：

```
    private void ctMenuOpenAttribute_Click(object sender, EventArgs e)
    {
        OpenAttributeTableFrm gridForm =
                    new OpenAttributeTableFrm (m _ tocRightLayer as
IFeatureLayer);
        gridForm.ShowDialog();
    }
```

按下 F5 键，编译运行程序，即可实现本章开篇处展示的属性表。

**(4)Shape 字段显示优化**

可以看出上述功能在显示 Shape 字段时不太美观，我们也可以像 ArcMap 一样显示字段的几何类型。首先用一个 getShapeTypeString(IFeatureClass pFCls)函数获取 Shape 字段的几何类型字符串，再利用 IFields 的 FindField()函数找到 Shape 字段的位置索引。

getShapeTypeString()函数代码如下：

```
    //获取 Shape 字段显式字符串
    private string getShapeTypeString(IFeatureClass pFCls)
    {
        string shape = "";
        if (pFCls.ShapeType == esriGeometryType.esriGeometryPoint)
            shape = "Point";
```

```
        else if (pFCls.ShapeType == esriGeometryType.esriGeometry
Polyline)
            shape = "Polyline";
        else if (pFCls.ShapeType == esriGeometryType.esriGeometry
Polygon)
            shape = "Polygon";

        return shape;
    }
```

修改 GenerateDataTable( )函数，在遍历要素前添加如下代码：

```
    string strShape = getShapeTypeString( pFCls);
        int ishape = pFCls.Fields.FindField("Shape");
```

对行元素赋值时，对字段索引等于“ishape”的列，赋值 strShape，其他的等于要素字段的值。修改代码如下：

```
    if (i == ishape)
        pRow[i] = strShape;
    else
        pRow[i] = pFea.get_Value(i).ToString();
```

## 五、功能增强

以上源代码只能实现全部要素的属性显示，下面通过一个按钮 btnViewMethod 实现选择集显示和全要素显示两种情形切换：

①在 Form 底部添加一个 Panel，然后在 Panel 上添加按钮 btnViewMethod，设置按钮的显示文本属性 Text = “Selected”。

②查询游标由函数 getSearchCursor( )获取，此函数根据布尔变量 isSelectionSet 的值决定是使用选择集创建游标，还是用要素创建游标，代码如下：

```
    ///<summary>
    ///从 Layer 查询到 Cursor
    ///</summary>
    ///<returns></returns>
    private ICursor getSearchCursor (IFeatureLayer pFLayer, bool
isSelectionSet)
    {
        ICursor pCursor = null;
        if (isSelectionSet)
        {
            IFeatureSelection pSeletion = pFLayer as IFeatureSelection;
```

```
            ISelectionSet pSelectionSet = pSeletion.SelectionSet;
            pSelectionSet.Search(null, false, out pCursor);
        }
        else
        {
            ITable pTable = pFLayer as ITable;
            pCursor = pTable.Search(null, false);
        }

        return pCursor;
    }
```

③修改 GenerateDataTable 函数参数，代码如下：

```
public DataTable GenerateDataTable ( IFeatueLayer pFealyr, bool isSelectSet )
```

将该函数创建查询游标的代码：

```
ICursor pCursor = (pFCls as ITable).Search(null, false);
```

改为：

```
ICursor pCursor = getSearchCursor( pFealyr, isSelectSet );
```

④添加 btnViewMethod 响应函数，点击按钮【DataGridView】，显示内容在选择集和全要素来回切换。代码如下：

```
private void btnViewMethod_Click(object sender, EventArgs e)
{
    if (btnViewMethod.Text == "Selected")
    {
        AttrGridView.DataSource = this.GenerateDataTable(m_pFeatureLayer, true);
        btnViewMethod.Text = "All";
    }
    else if (btnViewMethod.Text == "All")
    {
        AttrGridView.DataSource = this.GenerateDataTable(m_pFeatureLayer, false);
        btnViewMethod.Text = "Selected";
    }
}
```

## 六、编译测试

- ◆ 单击 F5 键编译运行程序；
- ◆ 添加数据：.. \\ Data \\ 制图数据 \\ 地级城市驻地 . shp；
- ◆ 右键单击【地级城市驻地】图层，选择图层操作浮动菜单【OpenAttributeTable】，打开“地级城市驻地”属性表。

效果如图 10-1 所示。

## 七、思考与练习

①如何为要素类添加一个新字段？

②如何新建字段统一赋值，例如【Length】字段，如何赋值几何长度？

③如何在 DataGridView 上选择要素？

# 实验十一　空间数据查询(基于属性)

## 一、目的与要求

①熟悉 IQueryFilter 接口建立属性约束条件的方法；
②熟悉 IFeatureSelection 进行空间查询的方法；
③熟悉查询结果集的合成方法；
④熟悉为要素选择显示配置颜色的方法。

## 二、实验原理

ArcEngine 中 QueryFilterClass 类是一个依据属性约束条件的查询过滤器，IQueryFilter 是该类实现的主要接口，通过对 IQueryFilter 的 WhereClause 属性设置任意复杂度的 SQL 条件子句，满足第一类查询条件(基于属性)的过滤要求。

ArcEngine 执行查询的方法如下：

①IFeatureClass 的 Search( ) 函数和 Select( ) 函数，前者返回查询结果游标(ICursor)，后者返回选择集(ISelectionSet)。对于游标，必须遍历游标才能得到所有的结果，不必关注内存；对于选择集，查询后即可得到，但是通常只保留 OID 结果集。

②图层类 FeatureLayerClass 有一个选择集属性(ISelectionSet)，该属性保存被选中的要素信息，而且还控制被选中的要素高亮显示。使用该类的 IFeatureSelection 接口的 SelectFeature( ) 函数可实现应用过滤器更新图层选择集，该功能是基于 IFeatureClass 的 Select( ) 函数扩展而成的。

③IFeatureClass 的 Search(IQueryFilter，bool) 函数有两个参数，一个是查询过滤器参数(IQueryFilter)，用于设置约束条件，另一个是布尔类型的回收参数(Recycling)，如果该参数为真，则在遍历查询游标的时候，所有要素将共享一个内存，也就是得到一新要素后，上一个要素的内存被回收。

## 三、实验环境

①开发环境：Visual Studio 2015 + ArcGIS Engine 10.5；
②开发语言：C#；
③实验数据：.. \\ Data \\ 制图数据 \\ (全国政区人口数据 shapefile)；

◆ 地级城市驻地.shp;
◆ 国界线.shp;
◆ 省级行政区.shp;
……

## 四、内容与步骤(基于属性查询)

功能描述:在 Query 菜单中,单击"Query By Attribute"菜单项,可弹出查询对话框,效果如图 11-1 所示。

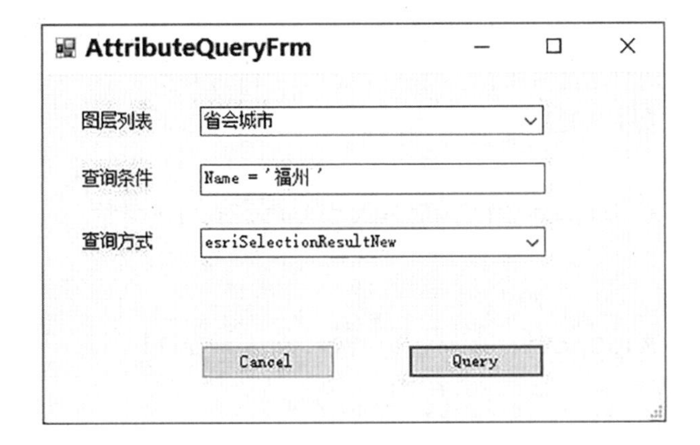


图 11-1　AttributeQueryFrm 控件布局

### 1. 查询窗体设计

**(1)创建窗体**

新建一个 Windows 窗体,命名为"QueryByAttributeFrm.cs",从工具箱中拖两个 ComboBox(图层列表,查询方式)、一个 TextBox(填写查询条件)、两个 Button(btnQuery、Cancel)控件到窗体。

表 11-1　**QueryByAttributeFrm 窗体控件命名表**

| 控件 | Name 属性 | 含义 |
| --- | --- | --- |
| ComboBox | cbxLayers | 选择图层 |
| ComboBox | cbxResultMethod | 结果方式 |
| TextBox | txtWhereClause | 查询条件 |
| Button | btnQueryByAttribute | 依属性查询 |
| Button | btnCancel | 取消 |

**(2)添加引用**

在解决方案资源管理器中添加 ArcGIS Engine 的 ERSI ArcGIS Geodatabase 等引用，在 QueryByAttributeFrm. cs 文件中添加如下 USing 指令：

```
using ESRI.ArcGIS.esriSystem;
using ESRI.ArcGIS.Carto;
using ESRI.ArcGIS.Geodatabase;
using ESRI.ArcGIS.Geometry;
using ESRI.ArcGIS.Controls;
```

**(3)添加功能函数**

- 添加两个私有成员：IMapControl3 m、mapControl；
- 修改构造函数参数 IMapControl3 mapControl，为私有成员赋值；
- 添加 Load 事件响应函数；
- 添加查询按钮 Click 事件响应函数。

代码如下：

```
public partial class QueryByAttributeFrm: Form
{
    //私有成员
    private IMapControl3 m_mapControl = null;
    //构造函数
    public QueryByAttributeFrm(IMapControl3 mapControl)
    {
        InitializeComponent();
        m_mapControl = mapControl;
    }

    //装载事件响应函数
     private void QueryByAttributeFrm _ Load ( object sender,
EventArgs e)
    //查询按钮响应函数
     private void btnQueryByAttributeBtn _Click ( object sender,
EventArgs e)
}
```

2. 功能实现

**(1) Load 响应函数实现**

窗体装载时完成两件事：一是用层名填充 cbxLayers，这里用到辅助函数 getLayers( )，它以枚举器形式返回 Map 中所有矢量数据图层接口集合；二是用五种方式查询结果类型字符串(New、Add、And、Xor、Subtract)填充 cbxResultMethod。代码如下：

```
//装载事件响应函数
private void QueryByAttributeFrm_Load(object sender, EventArgs e)
{
    //load all the feature layers in the map to the layers combo
    IEnumLayer layers = getLayers();
    layers.Reset();
    ILayer layer = null;
    while ((layer = layers.Next()) ! = null)
    {
        cbxLayers.Items.Add(layer.Name);
    }

    //loadall esri Selection Result Enum to the ResultMethod combo
    this.cbxResultMethod.Items.Add("New");
    this.cbxResultMethod.Items.Add("Add");
    this.cbxResultMethod.Items.Add("And");
    this.cbxResultMethod.Items.Add("Xor");
    this.cbxResultMethod.Items.Add("Subtract");
}
```

**(2)查询响应函数实现**

①获取要素选择接口：根据层名利用辅助函数 getFeatureLayer (...)，获取 IFeatureLayer 接口对象，再转换为 IFeatureSelection；

②创建 IQueryFilter 过滤器：初始化，并为 WhereClause 属性赋值；

③查询方式转换，将界面下拉框选择到的查询结果字符串转换为枚举类型：esriSelectionResultEnum；此处用到辅助函数 ResultStringToEnum(...)；

④使用 IFeatureSelection 的 SelectFeatures()函数执行查询；

⑤刷新选择集(必要时可为选择显示配置颜色)。

代码如下：

```
//执行查询
private voidbtnQueryByAttributeBtn_Click(object sender, EventArgs e)
{
    //获取图层,此处用到辅助函数 getFeatureLayer(string)
    IFeatureLayer pFeatureLayer =
              getFeatureLayer((string)cbxLayers.SelectedItem);
    if (null == pFeatureLayer)
      return;

    //获取选择集接口
     IFeatureSelection  pFeatureSelection  =  pFeatureLayer  as
```

```
IFeatureSelection;
        ISelectionSet pSelectionSet = pFeatureSelection.SelectionSet;

        //创建 IQueryFilter 过滤器
        IQueryFilter pQueryFilter = new QueryFilterClass();
        pQueryFilter.WhereClause = txtWhereClause.Text;

        //查询方式转换,此处用到辅助函数 ResultStringToEnum(string)
        esriSelectionResultEnum resultMethod;
        resultMethod = ResultStringToEnum((string)cbxResultMethod.
SelectedItem);

        //执行查询
        pFeatureSelection.SelectFeatures(pQueryFilter, resultMethod,
false);

        //为选择显示配置颜色
        //Color color = Color.FromArgb(255, 0, 0);
        //pFeatureSelection.SetSelectionSymbol=false;//否则颜色无效
        //pFeatureSelection.SelectionColor = ColorToIRgbColor(color);

        //选择集刷新
        this.m_mapControl.ActiveView.PartialRefresh(esriViewDraw
                                Phase.esriViewGeoSelection, null, null);
    }
```

**(3)辅助函数实现**

根据层类型 UID 获取矢量图层，代码如下：

```
    private IEnumLayer getLayers()
    {
        UID uid = new UIDClass();
        uid.Value = "{40A9E885-5533-11d0-98BE-00805F7CED21}";
        IEnumLayer layers = m_mapControl.Map.get_Layers(uid, true);
        return layers;
    }
```

根据层名获取矢量图层，代码如下：

```
    private IFeatureLayergetFeatureLayer(string layerName)
    {
        //get the layers from the maps
```

```
        IEnumLayer layers =getLayers();
        layers.Reset();
        ILayer layer = null;
        while ((layer = layers.Next()) ! = null)
        {
            if (layer.Name == layerName)
                return (layer as IFeatureLayer);
        }
        return null;
    }
```

转换查询方式字符为相应枚举类型，代码如下：

```
    private esriSelectionResultEnum ResultStringToEnum(string strMethod)
    {
        esriSelectionResultEnum result;
        switch( strMethod )
        {
            case "New":
              result = esriSelectionResultEnum.esriSelectionResultNew;
              break;
            case "Add":
              result = esriSelectionResultEnum.esriSelectionResultAdd;
              break;
            case "And":
              result = esriSelectionResultEnum.esriSelectionResultAnd;
              break;
            case "XOR":
              result = esriSelectionResultEnum.esriSelectionResultXOR;
              break;
            case "Subtract":
              result =esriSelectionResultEnum.esriSelectionResult
Subtract;
              break;
            default:
              result = esriSelectionResultEnum.esriSelectionResultNew;
              break;
        }
        return result;
    }
```

### 3. 调用查询窗体

在主菜单条上添加Query菜单，添加“Query By Attribute”菜单项，并建立Click事件响应函数，代码如下：

```
private void queryByAttrbuteToolStripMenuItem _ Click ( object
sender, EventArgs e)
{
    QueryByAttributeFrm queryfrm = new QueryByAttributeFrm( m_
mapContrl );
    Queryfrm.Show( );
}
```

## 五、编译测试

①单击F5键，编译运行程序。

②加载数据：… \\ Data \\ 制图数据 \\ 省会城市 . shp。

③点击菜单【Selecte By Attibute】，启动查询对话框，选择

- ◆ 待查询图层＝省会城市；
- ◆ 查询条件：Name ＝‘福州’；
- ◆ 结果合成方式＝NEW；

④点击【Query】按钮，查询结果将高亮显示。

效果如图11-2所示。

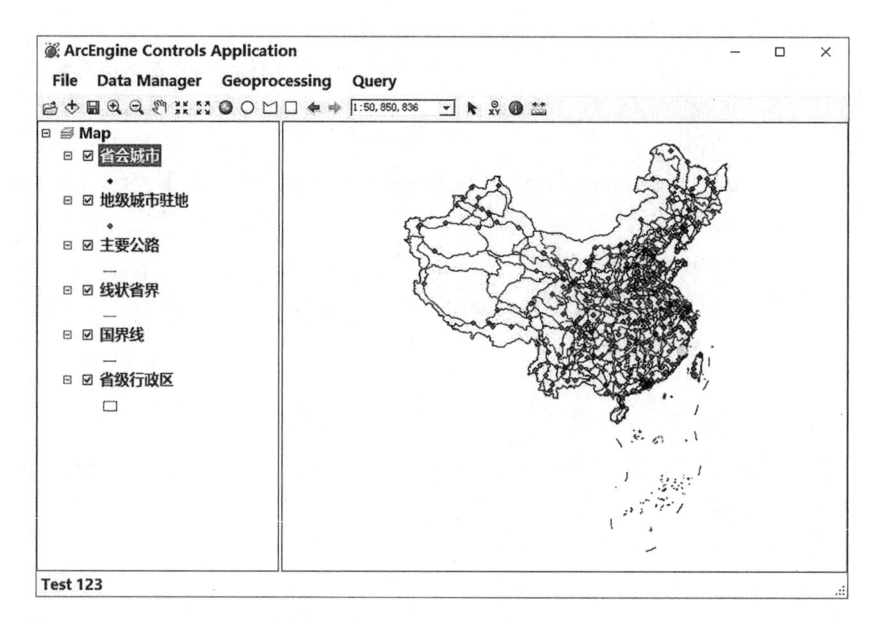


图11-2　基于属性查询效果图

## 六、思考与练习

①IQueryFilter 的查询字句是否支持所有的 SQL where 语法?
②能否像 SQL 查询一样指定返回的字段集?

# 实验十二 空间数据查询(基于空间关系)

## 一、目的与要求

①熟悉 ISpatialFilter 接口设置空间查询约束条件的基本方法;
②理解 esriSpatialRelEnum 空间关系类型的含义，及其在空间查询中的作用;
③熟悉使用 IGraphicsContainerSelect 接口操作图形元素的方法。

## 二、实验原理

ArcEngine 中 SpatialFilterClass 类是一个依据空间约束条件的查询过滤器，ISpatialFilter 是该类实现的主要接口，通过对 ISpatialFilter 的 Geometry 和 SpatialRel 属性设置，可满足基于空间关系查询的过滤要求：Geometry 用于设置查询的参考空间对象，SpatialRel 用于设置参考对象与目标对象之间的空间关系，包括相交、包含、重叠、相接、在内部等。

ISpatialFilter 继承了 IQueryFilter 接口，所以实际上具备了联合查询的能力，但从功能设计上都是分开设计的，是通过控制查询目标集达到联合查询的目的，例如，先进行基于属性的查询得到新选择集，然后以当前选择集执行基于空间关系的查询。

## 三、实验环境

①开发环境：Visual Studio 2015 + ArcGIS Engine 10. 5。
②开发语言：C#。
③实验数据：.. \\ Data \\ 制图数据 \\ (全国政区人口数据 shapefile):
◆ 地级城市驻地 . shp;
◆ 国界线 . shp;
◆ 省级行政区 . shp;
……

## 四、内容与步骤

功能描述：首先在视图中画一个闭合图形元素，并选定该元素，然后在 Query 菜单中，单击【Query By Geographic】菜单项，便可查询位于图形元素范围内的可见要素数据层，并高亮显示。

注意：为提供空间查询需要 IGeometry，应在工具条上添加绘制 GraphicElement 工具(例如：圆、矩形、多边形等)。

## 1. 创建功能类

**(1)添加引用**

```
using ESRI.ArcGIS.esriSystem;
using ESRI.ArcGIS.Carto;
using ESRI.ArcGIS.Geodatabase;
using ESRI.ArcGIS.Geometry;
using ESRI.ArcGIS.Controls;
```

**(2)添加功能函数**

①新建一个 C#类，并命名为"GeographicQueryClass"；

②添加两个私有成员：IMapControl3 m、mapControl ；

③修改构造函数参数：IMapControl3 mapControl;

④添加查询函数：QueryByGeographic()。

```
    public class GeographicQueryClass
    {
        //私有成员
        IMapControl3 m_mapControl = null;
        //构造函数
        public GeographicQueryClass(IMapControl3 mapControl)
        {
            m_mapControl = mapControl;
        }

        //查询功能函数
         private void QueryByGeographic( esriSelectionResultEnum
resultMethod)
        //辅助函数
        public IGeometry getGeometryRef()
        private IEnumLayer getLayers()
    }
```

## 2. 查询函数 QueryByGeographic()的实现

①利用 getLayers()函数获取 IEnumLayer 接口对象;

②创建 SpatialFilter 空间过滤器对象，这里使用 getGeometryRef()函数获取空间参考对象，为过滤器 Geometry 属性赋值;

③遍历所有图层，执行空间选择查询；

④刷新查询对象。

代码如下：

```
private void QueryByGeographic(esriSelectionResultEnum result
Method)
{
    //1:get the layers from the maps
    IEnumLayer layers =getLayers();
    layers.Reset();

    //2:创建 SpatialFilter 空间过滤器对象
    ISpatialFilter pSpatialFilter = new SpatialFilterClass();
    pSpatialFilter.Geometry = getGeometryRef(); //设置过滤器的
Geometry
    pSpatialFilter.SpatialRel = esriSpatialRelEnum.esriSpatial
RelIntersects; //设置空间关系类型

    //3:遍历所有图层,执行空间选择查询
    ILayer pLayer = null;
    while ((pLayer = layers.Next()) != null)
    {
        IFeatureSelection pFeatureSelection = pLayer as IFeature
Selection;
        //ISelectionSet pSelectionSet = pFeatureSelection.
SelectionSet;
        pFeatureSelection.SelectFeatures(pSpatialFilter,
resultMethod, false);
    }

    //4:刷新查询对象
     this.m_mapControl.ActiveView.PartialRefresh(esriViewDraw
Phase.esriViewGeoSelection, null, null);
}
```

### 3. 辅助函数

getGeometryRef()函数获取过滤器需要的空间参考对象。为简单起见，这里只取地图中第一个元素。

首先，利用 IGraphicsContainerSelect 接口的 SelectedElements 属性(它为图形元素枚举

器：IEnumElement)得到所有被选择的图形元素，然后以第一个元素的几何图形为基础创建一个0.1个单位的缓冲区，作为查询的空间限制条件。

代码如下：

```
public IGeometry getGeometryRef()
{
    IGraphicsContainerSelect pContainerSelect = null;
    pContainerSelect = m_mapControl.Map as IGraphics
ContainerSelect;
    pContainerSelect.SelectAllElements();
    IEnumElement pEnumGraphics = pContainerSelect.
SelectedElements;

    IGeometry pBuffer = null;
    IElement pElement = pEnumGraphics.Next();
    if (pElement ! = null)
    {
        //计算0.1单位缓冲区：
        ITopologicalOperator pTopo = null;
        pTopo = (pElement.Geometry) as ITopologicalOperator;
        pBuffer = pTopo.Buffer(0.1);
    }

    return pBuffer;
}
```

用到的辅助函数getLayers()代码如下：

```
//根据层类型UID获取矢量图层
private IEnumLayer getLayers()
{
     UID uid = new UIDClass();
     uid.Value = "{40A9E885-5533-11d0-98BE-00805F7CED21}";
     IEnumLayer layers = m_mapControl.Map.get_Layers(uid, true);
     return layers;
}
```

## 4. 调用查询窗体

①在主菜单条上添加Query菜单，添加【Query By Geographic】菜单项，并建立Click事件响应函数；

②最后，调用 GeographicQueryClass 的 QueryByGeographic(…)函数。

代码如下：

```
private void queryByGraphicsToolStripMenuItem _ Click ( object
sender, EventArgs e)
{
    //调用查询类
    GeographicQueryClass pQuery = null;
    pQuery = new GeographicQueryClass(m_mapControl );
    pQuery .QueryByGeographic(esriSelectionResultEnum.
esriSelectionResultNew);
}
```

## 五、编译测试

①单击 F5 键，编译运行程序；

②添加数据："… \\ Data \\ 制图数据 \\ "目录下所有文件；

③用图形元素绘制工具(New Circle、New Regtangle、New Polygon 等)在地图有效区域绘制若干图形元素(例如：圆形)；

④点击菜单【Query By Geographic】进行查询。

效果如图 12-1 所示。

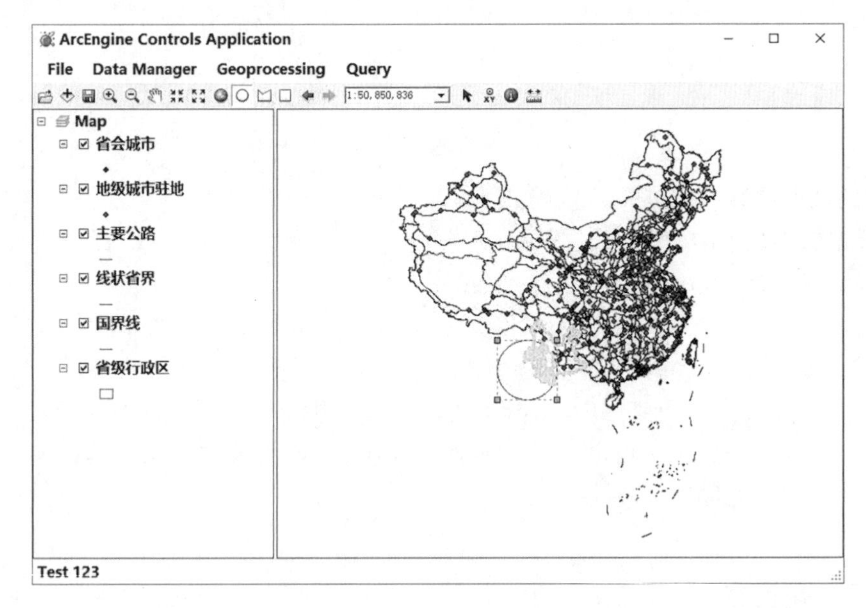


图 12-1　基于空间关系查询效果图

## 六、思考与练习

①修改空间关系类型(例如：包含、在内部)，查看分析效果。

②修改空间约束，使之支持多个图形的空间查询。

③功能完成后，可以配合操作。例如：先进行空间查询，打开属性表查看选择集中的要素集。然后，执行基于属性查询，再打开属性表查看选择集中的要素集。

# 实验十三　统 计 计 算

## 一、目的与要求

熟悉使用 DataStatistics 组件对要素类进行统计计算的方法。

## 二、实验原理

ArcGIS Engine 提供了一般性统计方法：

①Count——个数；

②Maximum——最大值；

③Mean——算术平均值；

④Minimum——最小值；

⑤StandardDeviation——标准差；

⑥Sum——求和。

统计结果包括：

①BaseStatistics（基础统计）组件用来生成和报告任意数值集合的统计结果。其中 IFrequencyStatistics 接口提供用来报告频率统计的结果，IGenerateStatistics 接口提供生成统计结果数据的功能，IStatisticsResults 接口提供对报告各统计结果的功能。

②DataStatistics 组件提供对单个字段的统计计算及单个字段的唯一值。组件创建后，用来分析的数据通过 IDataStatistics 接口的 Cursor 属性，以游标的形式传入输入表，注意 ICursor 的对象只能使用一次，如果要获取多个结果，应当再次创建游标。

IDataStatistics 是 DataStatistics 统计组件实现的唯一接口，IDataStatistics 属性包括：

①Cursor——通过游标传递输入表；

②Field——待统计的字段；

③UniqueValueCount——统计表中唯一值总数；

④UniqueValues——唯一值枚举 ；

⑤Statistics——IStatisticsResults 对象，用于返回统计信息。

## 三、实验环境

①开发环境：Visual Studio 2005。

②开发语言：C#。

③实验数据：…　\\ Data \\ 制图数据：

◆　地级城市驻地 . shp；

◆　国界线 . shp；

◆　省级行政区 . shp；

◆　主要公里(分政区). shp，其中，字段【NAME】为政区名称，字段【Shape_Length】为公路里程。

## 四、内容与步骤

本实验实现统计计算的基本功能：用户右键单击浮动菜单上 Statistics 菜单项，激活统计计算对话框，用户可选择分组字段、统计字段、统计方法等。

### 1. StatisticsFrm 窗体设计

#### (1)添加统计计算对话框类

新建一个 Windows 窗体，命名为“StatisticsFrm”，修改窗体的 Text 属性为 StatisticsFrm，并添加 Button、Label、TextBox、ComboBox、DataGridView 控件。控件布局如图 13-1 所示。

StatisticsFrm

GroupField: NAME　StaticField: Shape_Leng　Method: SUM

| NAME | Frequency | SUM_Shape_Leng |
| --- | --- | --- |
|  | 9 | 78701.040536738 |
| 内蒙古 | 78 | 5752607.50729269 |
| 黑龙江 | 51 | 3291888.5571676 |
| 吉林 | 43 | 2659434.23099322 |
| 辽宁 | 61 | 3245733.3758752 |
| 宁夏 | 25 | 912610.90516378 |
| 陕西 | 82 | 3454113.4185586 |
| 山西 | 95 | 3311893.16062218 |
| 河北 | 121 | 4535111.56396281 |
| 山东 | 114 | 4545036.82074682 |
| 甘肃 | 76 | 4475261.63847317 |
| 北京 | 19 | 817585.67713353 |
| 天津 | 12 | 394695.3679739 |
| 河南 | 85 | 3528991.56519548 |

Save As Dbf　Cancel　App

图 13-1　StatisticsFrm 窗体控件布局

### (2)设置控件属性

设置相应控件的相关属性，见表 13-1。

表 13-1 **StatisticsFrm 窗体控件命名表**

| 控件类型 | Name 属性 | 含义 | 备注 |
|---|---|---|---|
| ComboBox | cbxGroupField | 分组字段名 | 字符型字段 |
| ComboBox | cbxStatisticsField | 统计字段名 | 数值型字段 |
| ComboBox | cbxMethod | 统计方法 | |
| DataGridView | dataGridView1 | 统计结果列表 | |
| Button | btnBrower | 设置输出文件 | |
| Button | btnApp | 应用 | |
| Button | btnCancel | 取消 | |
| Button | btnSaveAsDbf | 结果保存为 Dbf | |

### (3)添加 StatisticsFrm 的全局变量

```
IFeatureLayer _pFeatureLayer = null;
```

### (4)添加 StatisticsFrm 事件响应函数、功能函数、辅助函数

①添加应用按钮 Click 事件响应函数；

②添加确定按钮 Click 事件响应函数。

具体代码如下：

```
public partial class StatisticsFrm : Form
{
    IFeatureLayer _pFeatureLayer = null;
    public StatisticsFrm(IFeatureLayer featuerLayer )
    {
        InitializeComponent();
        _pFeatureLayer = featuerLayer;
    }

    //事件响应函数
    private void StatisticsFrm_Load(object sender, EventArgs e)
    private void btnOK_Click(object sender, EventArgs e)
    private void btnCancel_Click(object sender, EventArgs e)
    private void btnSaveAsDbf_Click(object sender, EventArgs e)

    //功能函数
```

```
        public DataTable Aggregate(string sumField, string Method)
        public DataTable Aggregate(string groupField, string sumField, string Method)
        public DataTable AggregateForGeodatabase(string groupField, string sumField, string Method)

        //辅助函数
        private IEnumerator GetUniqueValues(IFeatureClass pFC, string groupField)
        private string ConvertStaticString(IStatisticsResults pResults, string Method)
    }
```

2. StatisticsFrm 类的实现

**(1)Load 事件响应函数的实现**

对选定的图层要素类字段集，分别用字符型字段和数值型字段填充分组字段和统计字段控件(cbxGroupField、cbxStaticField)，统计操作符(“SUM”，“AVG”等)填充 cbxMethod 控件，代码如下：

```
private void StatisticsFrm_Load(object sender, EventArgs e)
{
    IFields pFields = _pFeatureLayer.FeatureClass.Fields;
    this.cbxGroupField.Items.Add("");
    for (int k = 0; k <pFields.FieldCount; k++)
    {
        IField pFd = pFields.get_Field(k);
        switch (pFd.Type)
        {
            //分组字段只支持字符型
            case esriFieldType.esriFieldTypeString:
                this.cbxGroupField.Items.Add(pFd.Name);
                break;
            case esriFieldType.esriFieldTypeDouble:
            case esriFieldType.esriFieldTypeInteger:
            case esriFieldType.esriFieldTypeSingle:
            case esriFieldType.esriFieldTypeOID:
                this.cbxStaticField.Items.Add(pFd.Name);
                break;
            default:
```

```
                break;
            }
        }

        this.cbxMethod.Items.Add("SUM");
        this.cbxMethod.Items.Add("MIN");
        this.cbxMethod.Items.Add("MAX");
        this.cbxMethod.Items.Add("AVG");
        this.cbxMethod.Items.Add("STDDEV");

        this.cbxGroupField.SelectedIndex = 0;
        this.cbxStaticField.SelectedIndex = 0;
        this.cbxMethod.SelectedIndex = 0;
    }
```

**(2)应用响应函数 btnApp_Click()**

应用响应函数根据分组字段取值是否有效，决定调用分组聚合函数，还是一般聚合函数。统计结果存放在 DataTable 中，作为 DataGridView 的数据源，在表格空间中显示出来。

代码如下：

```
    private void btnOK_Click(object sender, EventArgs e)
    {
        string groupField = cbxGroupField.SelectedItem.ToString();
        string sumField = cbxStaticField.SelectedItem.ToString();
        string Method = cbxMethod.SelectedItem.ToString();

        DataTable dt = null;
        if(groupField ! ="")
            //dt = AggregateForGeodatabase(groupField, sumField,
Method);
            dt = Aggregate(groupField, sumField, Method);
        else
            dt = Aggregate(sumField, Method);

        this.dataGridView1.DataSource = dt;
    }

    private void btnCancel_Click(object sender, EventArgs e)
    {
        this.Close();
```

```
}
```

**(3) btnSaveAsDbf_Click( )函数**

本函数调用Dbf操作类DbfOper的WriteDbf( )函数，将DataGridView的数据源中数据写为Dbf表。结果存为当前目录。DbfOper源代码参看附录2。

```
private void btnSaveAsDbf_Click(object sender, EventArgs e)
{
    DbfOper writeDbf = new DbfOper(System.Environment.
CurrentDirectory);
    writeDbf.WriteDbf(this.dataGridView1.DataSource as DataTable);
}
```

**(4)核心函数的实现**

Aggregate( )不分组聚合函数，实现步骤如下：

第一步，初始化DataTable，用频数(Frequency)、“统计方法”+“字段名”，建立表头。

第二步，使用IDataStatistics获取统计结果；

第三步，将结果添加为DataTable的一行。

代码如下：

```
public DataTable Aggregate(string sumField, string Method)
{
    string frquencyField = "Frequency";
    string staticField = Method + "_" + sumField;

    //初始化DataTable
    DataTable pTable = new DataTable(_pFeatureLayer.Name);
    pTable.Columns.Add(frquencyField);
    pTable.Columns.Add(staticField);

    //获取统计结果
    IFeatureClass pFC = _pFeatureLayer.FeatureClass;
    ICursor groupCursor = pFC.Search(null, false) as ICursor;
    IDataStatistics pDatdaS = new DataStatisticsClass();
    {
        pDatdaS.Field = sumField;
        pDatdaS.Cursor = groupCursor;
    }
    IStatisticsResults pResults = pDatdaS.Statistics;

    //将结果添加为DataTable的一行
    DataRow pRow = pTable.NewRow();
```

```
        pRow[frquencyField] = pResults.Count;
        pRow[staticField] = ConvertStaticString(pResults, Method);
        pTable.Rows.Add(pRow);

        return pTable;
    }
```

Aggregate()分组聚合函数，实现步骤如下：

第一步，初始化 DataTable，用“分组字段名”、频数(Frequency)、“统计方法”+“字段名”，建立表头。

第二步，获取分组字段的唯一值集合，使用辅组函数 GetUniqueValues()。

第三步，遍历唯一值集合，为分组字段每个唯一值，在 DataTable 中生成一行统计数据：

①对每个唯一值构造一个查询过滤器；

②用过滤器获取等于该唯一值的游标；

③用 IDataStatistics 获取统计结果；

④将结果添加为 DataTable 的一行。

代码如下：

```
public DataTable Aggregate(string groupField, string sumField, string Method)
    {
        string frquencyField = "Frequency";
        string staticField = Method + "_" + sumField;

        DataTable pTable = new DataTable(_pFeatureLayer.Name);
        pTable.Columns.Add(groupField);
        pTable.Columns.Add(frquencyField);
        pTable.Columns.Add(staticField);

        //获取分组字段的唯一值集合
        IFeatureClass pFC = _pFeatureLayer.FeatureClass;
        IEnumerator Em = GetUniqueValues(pFC, groupField);
        Em.Reset();
        while (Em.MoveNext())
        {
            object obj = Em.Current;

            //按分组条件构造过滤器
            IQueryFilter pQueryFilter = new QueryFilterClass();
```

```
            pQueryFilter.WhereClause = groupField + "='"+ obj.
ToString()+"'";
            ICursor groupCursor = pFC.Search(pQueryFilter, false) as
ICursor;

            //获取统计结果
            IDataStatistics pDatdaS = new DataStatisticsClass();
            {
                pDatdaS.Field = sumField;
                pDatdaS.Cursor = groupCursor;
            }
            IStatisticsResults pResults = pDatdaS.Statistics;

            //将结果在Table中添加行
            DataRow pRow = pTable.NewRow();
            pRow[groupField] = obj;
            pRow[frquencyField] = pResults.Count;
            pRow[staticField] = ConvertStaticString(pResults, Method);
            pTable.Rows.Add(pRow);
        }

        return pTable;
    }
```

对于数据源为数据库的分组聚合计算，可用工作空间创建IQueryDef2接口，然后调用IQueryDef2的Evaluate2()方法，直接获得分组聚合结果，此方法需要为IQueryDef2接口配置聚合SQL语句，这就利用了数据库聚合运算功能，计算速度会快很多。参考代码如下：

```
    public DataTable AggregateForGeodatabase (string groupField,
string sumField, string Method)
    {
        string frquencyField = "Frequency";
        string staticField = Method + "_" + sumField;

        DataTable pTable = new DataTable(_pFeatureLayer.Name);
        pTable.Columns.Add(groupField);
        pTable.Columns.Add(frquencyField);
        pTable.Columns.Add(staticField);
```

```
        IDataset pDataset = _pFeatureLayer.FeatureClass as IDataset;
        IFeatureWorkspace pFWorkspace = pDataset.Workspace as
IFeatureWorkspace;
        IQueryDef2 qf = pFWorkspace.CreateQueryDef() as IQueryDef2;
        qf.Tables = _pFeatureLayer.Name;
        qf.SubFields = groupField + ", COUNT(" + sumField + ") AS " +
frquencyField+", "+Method+ "(" + sumField + ") AS " + staticField;
        //qf.WhereClause = groupField + "<>"";
        //qf.WhereClause = groupField +" IS NOT NULL ";
        qf.PostfixClause = "GROUP BY " + groupField;
        ICursor feacur = qf.Evaluate2(false);

        //将结果在 Table 中添加行
        IRow pRow = null;
        int groupIDx = feacur.FindField(groupField);
        int frequencyIDx = feacur.FindField(frquencyField);
        int staticIDx = feacur.FindField(staticField);
        while ((pRow = feacur.NextRow())! = null)
        {
            DataRow pDataRow = pTable.NewRow();
            string str0 = pRow.get_Value(groupIDx).ToString();
            string str1 = pRow.get_Value(frequencyIDx).ToString();
            string str2 = pRow.get_Value(staticIDx).ToString();

            pDataRow[groupField] = str0;
            pDataRow[frquencyField] = str1;
            pDataRow[staticField] = str2;
            pTable.Rows.Add(pDataRow);
        }

        return pTable;
    }
```

**(5)辅助函数的实现**

GetUniqueValues()函数通过 IDataStatistics 实现，代码如下：

```
    private IEnumerator GetUniqueValues (IFeatureClass pFC, string
groupField)
    {
        ICursor cursor = pFC.Search(null, false) as ICursor;
```

```
        IDataStatistics pDatdaS = new DataStatisticsClass();
        {
            pDatdaS.Field = groupField;
            pDatdaS.Cursor = cursor;
        }

        IEnumerator Em = pDatdaS.UniqueValues;
        return Em;
    }
```

统计操作符转换函数代码如下:

```
    private string ConvertStaticString(IStatisticsResults pResults,
string Method)
    {
        string OutStr = null;
        switch (Method)
        {
            case "MIN":
                OutStr = pResults.Minimum.ToString();
                break;
            case "MAX":
                OutStr = pResults.Maximum.ToString();
                break;
            case "AVG":
                OutStr = pResults.Mean.ToString();
                break;
            case "STDDEV":
                OutStr = pResults.StandardDeviation.ToString();
                break;
            case "SUM":
            default:
                OutStr = pResults.Sum.ToString();
                break;
        }

        return OutStr;
    }
```

### 3. 功能调用

在图层操作浮动菜单上添加一菜单项，命名为“Statistics”，创建并修改 Click 事件响应函数，代码如下：

```
private void statisticsToolStripMenuItem_Click(object sender, EventArgs e)
{
    StatisticsFrm frm = new StatisticsFrm(this.m_tocRightLayer as IFeatureLayer);
    frm.Show();
}
```

## 五、功能测试

①单击 F5 键，编译运行程序；

②添加数据：... \\ Data \\ 制图数据 \\ 主要公里(分政区).shp；

③点击图层操作浮动菜单【Statistics】，弹出分析窗口，选择：

◆ 分组字段=【NAME】；

◆ 统计字段=【Shape_Length】；

◆ 统计方法= SUM；

④单击【App】铵钮即可生成统计结果。

效果如前文图 13-1 所示。

## 六、思考与练习

①对于数据源为数据库的分组聚合计算，可用 IQueryDef2 接口提高运行效率。试用 IQueryDef2 接口实现面向 GDB 数据库的分组查询统计。

②IFilterQuery 接口的 SubFields 属性起什么作用？

# 实验十四　要素融合

## 一、目的与要求

①熟悉采用 IBasicGeoprocessor 实现要素融合的方法；

②熟悉名称对象的构造方法。

## 二、实验原理

要素融合是将具有相同类别的要素合并为一个新的要素，例如：具有相同属性的两个相邻(或有重叠)多边形将合并成一个多边形；或者具有相同属性的两条相邻弧段将合并成一条弧段。图 14-1 为空间要素融合输出结果的示意图。

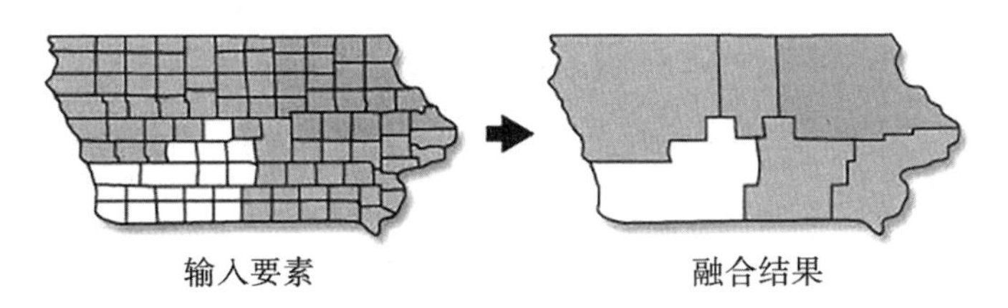


图 14-1　空间要素融合示意图

ArcGIS Engine 实现要素融合可使用 IBasicGeoprocessor 接口(BasicGeoprocessor 实现了本接口)的 Dissolve 方法，或使用 Geoprossing 的 Dissolve 工具。

## 三、实验环境

①开发环境：Visual Studio 2015 + ArcGIS Engine 10. 5。

②开发语言：C#。

③实验数据：... \\ Data \\ 矢量数据 \\ Ex4. mdb：

- 云南县界；
- Grid_Poly；

……

## 四、内容与步骤

本实验目的在于实现要素融合基本功能：用户右击【Geoprocessing】主菜单上【Dissolve】菜单项，激活要素融合对话框，用户可选择输入图层，输出文件，融合字段等。

1. Dissolve 窗体设计

**(1) 添加要素融合对话框类**

新建一个 Windows 窗体，并命名为“DissolveFrm”，修改窗体的 Text 属性为“Dissolve”，并添加 Button、Label、TextBox，ComboBox 控件。控件布局如图 14-2 所示。

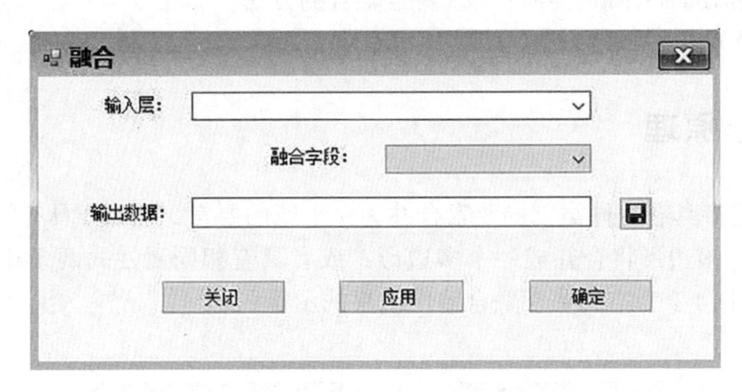


图 14-2　DissolveFrm 控件布局设计

**(2) 设置控件属性**

设置相应控件的相关属性，见表 14-1。

表 14-1　**DissolveFrm 控件命名表**

| 控件 | Name 属性 | 含义 |
|---|---|---|
| ComboBox | cbxInputLyr | 输入图层 |
| ComboBox | cbxDissolveField | 融合字段 |
| TextBox | txtOutPath | 输出路径 |
| Button | btnBrowserPath | 输出路径浏览 |
| Button | btnApp | 执行分析 |
| Button | btnCancel | 关闭 |
| Button | btnOK | 确定 |

**(3) 添加 DissolveFrm 的全局变量**

代码如下：

```
private IMapControl3 _mapControl;
```

**(4)添加 DissolveFrm 事件响应函数、功能函数、辅助函数**

①添加应用按钮 Click 事件响应函数;

②添加确定按钮 Click 事件响应函数。

代码如下:

```
public partial class DissolveFrm : Form
{
    private IMapControl3 _mapControl = null;
    public DissolveFrm(IMapControl3 mapControl)
    {
        InitializeComponent();
        _mapControl = mapControl;
    }
    //事件响应函数
    private void DissolveFrm_Load(object sender, EventArgs e)
    private void cbxInputLyr_SelectedIndexChanged(object sender,
EventArgs e)
    private void btnBrowserPath_Click(object sender, EventArgs e)
    private void btnApp_Click(object sender, EventArgs e)
    private void btnCancel_Click(object sender, EventArgs e)
    private void btnOK_Click(object sender, EventArgs e)

    //核心功能函数
     public IFeatureClass Dissolve (IFeatureLayer inFeatLayer,
string dissolveField, string pathName, string dsName)
    //辅助函数
     private IFeatureClassName CreateFeatureClassName (string
pathName, string dsName, esriGeometryType geoType)
    private ILayer GetLayerByname(string lyrName)
}
```

## 2. DissolveFrm 类的实现

### (1)Load 事件响应函数的实现

Load 事件响应函数的主要作用是:用 MapControl 中图层名填充 cbxInputLyr(输入图层)控件,代码如下:

```
private void DissolveFrm_Load(object sender, EventArgs e)
{
    for (int i = 0; i < _mapControl.Map.LayerCount; i++)
```

```
        {
            ILayer aLayer = _mapControl.Map.get_Layer(i);
            if (aLayer is IFeatureLayer)
            {
                IFeatureLayer flyr = (IFeatureLayer)aLayer;
                if (flyr.FeatureClass.ShapeType ==
                              esriGeometryType.esriGeometryPolygon)
                    this.cbxInputLyr.Items.Add(aLayer.Name);
            }
        }

        txtOutPath.Text = "";
        cbxDissolveField.Text = "无";
    }
```

**(2) btnBrowserPath 按钮响应函数的实现**

实现代码如下:

```
    private void btnBrowserPath_Click(object sender, EventArgs e)
    {
        if (this.folderBrowserDialog1.ShowDialog() == DialogResult.
OK)
        {
            string path = this.folderBrowserDialog1.SelectedPath;
            this.txtOutPath.Text = path;
        }
    }
```

**(3) 输入图层选择事件响应函数的实现**

本函数根据选择的图层，将符合融合操作要求的字段名，填充控件 cbxDissolveField。

```
    private void cbxInputLyr_SelectedIndexChanged(object sender,
EventArgs e)
    {
        IFeatureLayer pFeatLyr = GetLayerByname(cbxInputLyr.Text) as
IFeatureLayer;
        if (pFeatLyr == null)
            return;

        cbxDissolveField.Items.Clear();
        cbxDissolveField.Items.Add("无");
        IFields pFields = pFeatLyr.FeatureClass.Fields;
```

```
        for (int j = 0; j < pFields.FieldCount; j++)
        {
            IField pFiled = pFields.get_Field(j);
            switch (pFiled.Type)
            {
                case esriFieldType.esriFieldTypeDouble:
                case esriFieldType.esriFieldTypeSingle:
                case esriFieldType.esriFieldTypeSmallInteger:
                case esriFieldType.esriFieldTypeInteger:
                case esriFieldType.esriFieldTypeString:
                    cbxDissolveField.Items.Add(pFiled.Name);
                    break;
                default:
                    break;
            }
        }
    }
```

**(4) btnApp 按钮响应函数的实现**

btnApp 按钮响应函数可实现要素融合功能，步骤如下：

第一步，从界面获取分析参数：融合字段名，输出目录，输出文件名(这里程序自动生成文件名)等。

第二步，执行融合操作，利用核心函数 Dissolve()。

第三步，将结果添加到地图。

代码如下：

```
    private void btnApp_Click(object sender, EventArgs e)
    {
         IFeatureLayer pFLyr = GetLayerByname(cbxInputLyr.Text) as
IFeatureLayer;
        string dissloveField = this.cbxDissolveField.Text;
        string pathName = this.txtOutPath.Text;
        if(pathName == "")pathName=System.Windows.Forms.Application.
StartupPath;
        string dsName = pFLyr.Name + "_Dissolve" +
                    DateTime.Now.ToString("yyyy-MM-dd_HH-mm-ss");

        //执行融合计算
        IFeatureClass pFcls = Dissolve(pFLyr, dissloveField, pathName,
dsName);
```

```
        //结果添加到地图
        IFeatureLayer newFlyr = new FeatureLayerClass();
        newFlyr.FeatureClass = pFcls;
        newFlyr.Name = pFcls.AliasName;
        _mapControl.AddLayer(newFlyr, 0);
    }

    private void btnCancel_Click(object sender, EventArgs e)
    {
        this.Close();
    }

    private void btnOK_Click(object sender, EventArgs e)
    {
        this.Close();
    }
```

**(5)核心功能函数的实现**

核心功能函数可实现输入要素类融合，结果返回多边形要素：

第一步，创建输出要素名称对象 IDatasetName，因为 IBasicGeoprocessor 接口需要 IDatasetName 接口作为输出参数。

第二步，执行融合操作，包括初始化 IBasicGeoprocessor 接口变量，调用 Dissolve 函数执行；由于融合的结果往往是多个要素形成一个新要素，因此结果要素集的属性被设计为汇总字段，也就是 Dissolve 需要汇总字段集字符串(用逗号分格)作为参数，格式为：

"Dissolve. Shape, Minimum. SUB_REGION, Count. SUB_REGION, Average. AREA";

每个汇总字段构成："汇总操作符"."字段名"，其中"Dissolve"是图形汇总操作符，由该操作符构成的字段将为结果要素的 Shape 字段，如果结果中需要融合字段，可用 Minimum 操作符和融合字段构成相应的汇总字段。

第三步，结果转换为要素类。

代码如下：

```
    public IFeatureClass Dissolve(IFeatureLayer inFeatLayer, string
dissolveField, string pathName, string dsName)
    {
         // Use the Itable interface from the Layer (not from the
FeatureClass)
        ITable pInputTable = default(ITable);
        pInputTable = inFeatLayer as ITable;
```

```
        //创建 IDatasetName
        esriGeometryType type = inFeatLayer.FeatureClass.ShapeType;
        IDatasetName pDatasetName = CreateFeatureClassName(pathName,
dsName, type) as IDatasetName;

        //执行融合操作
        IBasicGeoprocessor iBGP = new BasicGeoprocessor();
        ITable pOutputTable = default(ITable);
        string summaryField = "Dissolve.Shape,Minimum."+
dissolveField;
        pOutputTable = iBGP.Dissolve(pInputTable, false, dissolveField,
summaryField,pDatasetName);

        //结果转换为要素类
        IFeatureClass pOutputFeatClass = pOutputTable as IFeatureClass;
        return pOutputFeatClass;
    }
```

**(6)辅助函数**

函数 CreateFeatureClassName()可创建一个要素类的名称对象(IFeatureClassName)，所有的名称(包括 IWorkspaceName、IDatasetName 等)对象都是 IName 的派生类。IName 对象是一个代表性对象。通过使用 IName 对象，可以访问它所代表的对象的一些基本属性，而不用将整个对象调入内存，这样可以节省资源，提高程序运行效率。例如，我们可以用 IWorkspaceName 对象获得工作空间的基本属性，不需要打开工作空间的全部内容(除非使用了 IWorkspaceName. Open)，如果使用 Factory 打开一个 IWorkspace，那是要调入内存的。

辅助函数 GetLayerByname()根据层名获得数据层接口。代码如下：

```
    private    IFeatureClassName    CreateFeatureClassName    ( string
pathName, string dsName, esriGeometryType geoType)
    {
        IFeatureClassName ftClassName = new FeatureClassNameClass();
        {
            ftClassName.FeatureType = esriFeatureType.esriFTSimple;
            ftClassName.ShapeFieldName = "Shape";
            ftClassName.ShapeType = geoType;
        }

        //工作空间名称
        IWorkspaceName pNewWSName = new WorkspaceNameClass();
        {
```

```
            pNewWSName.WorkspaceFactoryProgID =
                "esriDataSourcesFile.ShapefileWorkspaceFactory";
            pNewWSName.PathName = pathName;
        }

        //数据集名称
        IDatasetName pDatasetName = ftClassName as IDatasetName;
        {
            pDatasetName.Name = dsName;
            pDatasetName.WorkspaceName = pNewWSName;
        }

        return ftClassName;
    }

    private ILayer GetLayerByname(string lyrName)
    {
        ILayer pLayer = null;
        for (int i = 0; i < _mapControl.LayerCount; i++)
        {
            ILayer tempLayer = _mapControl.get_Layer(i);
            if (tempLayer.Name == lyrName)
            {
                pLayer = tempLayer;
                break;
            }
        }

        return pLayer;
    }
```

3. 功能调用

在主菜单 Geoprocessing 上添加一菜单项(命名为"Dissolve"),创建并修改 Click 事件响应函数,代码如下:

```
    private void dissolveToolStripMenuItem_Click_1(object sender,
EventArgs e)
    {
        DissolveFrm frm = new DissolveFrm(m_mapControl);
```

```
    if (frm.ShowDialog() = = DialogResult.OK)
    {
        this.m_mapControl.ActiveView.Refresh();
    }
}
```

## 五、功能测试

①单击 F5 键，编译运行程序；
②加载数据：… \\ Data \\ 矢量数据 \\ Ex4. mdb \\ 云南县界；
③点击菜单【Dissolve】启动要素融合对话框，设置输出路径，选择：
- ◆ 输入图层＝ 云南县界；
- ◆ 融合字段＝【所属州】；

④点击【应用】按钮，即可看到结果。
效果如图 14-3 所示。

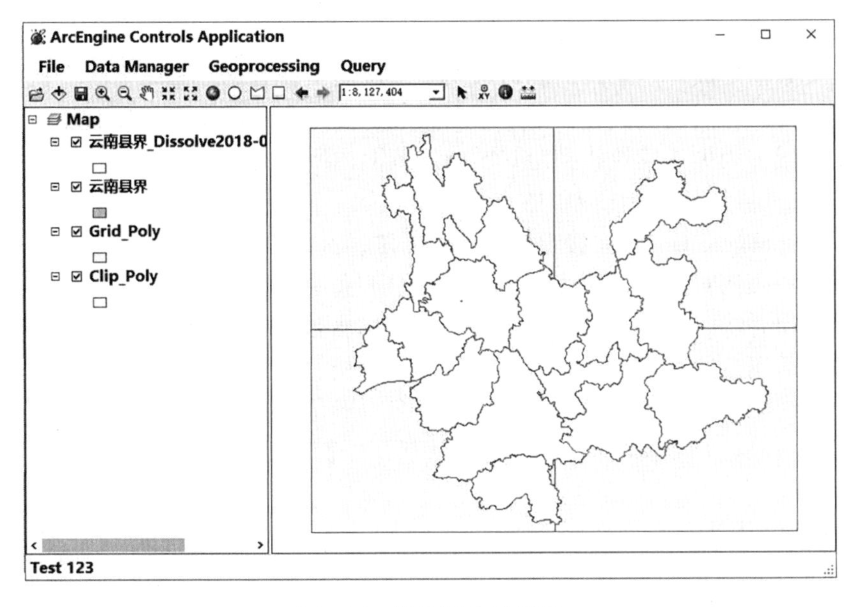


图 14-3　数据融合效果图

# 实验十五　缓冲区分析

## 一、目的与要求

①掌握使用 IBufferConstruction 接口进行缓冲区分析的步骤；
②理解缓冲多边形融合、分散的含义；
③掌握 IGeometryCollection 几何对象集合的用法。

## 二、实验原理

缓冲区分析(Buffer)是对选中的一组或一类地图要素(点、线或面)按设定的距离条件，围绕其要素而形成一定距离的多边形实体，从而实现数据在二维空间得以扩展的信息分析方法。缓冲区应用的实例有：污染源对其周围的污染量随距离增加而减小，从而可确定被污染的区域；为失火建筑找到距其 500 米范围内所有的消防水管等。

ArcGIS Engine 实现缓冲区分析的方法主要有如下三种：

①使用 ITopologicalOperator 接口的 Buffer 方法，高级几何对象都可实现这个接口，它适用于针对单一几何对象构建缓冲区操作，适用于底层操作。

②使用 IBufferConstruction 接口 ConstructBuffers 方法，BufferConstructionClass 可实现该接口，该方法适用于针对几何对象集合进行缓冲分析，可通过 IBufferConstructionProperties 接口控制生成缓冲多边形的过程，例如，重叠多边形是否融合等。

③使用 Geoprosser 的 Buffer 工具，此方法适用于面向要素图层的缓冲区操作。

## 三、实验环境

①开发环境：Visual Studio 2015 + ArcGIS Engine 10.5。
②开发语言：C#。
③实验数据：… \\ Data \\ 矢量数据 \\ 房屋道路：

- FW. shp；
- RC. shp；

……

## 四、内容与步骤

本实验目的在于实现缓冲区分析基本功能：用户右击【Geoprocessing】主菜单上【Buffer】菜单项，激活缓冲区分析对话框，用户可选择输入图层，输出文件以及控制参数表等。

### 1. Buffer 窗体设计

#### (1) 添加缓冲分析对话框类

新建一个 Windows 窗体，命名为“BufferFrm”，修改窗体的 Text 属性为“Buffer”，并且添加 Button、Label、ComboBox 控件。控件布局如图 15-1 所示。

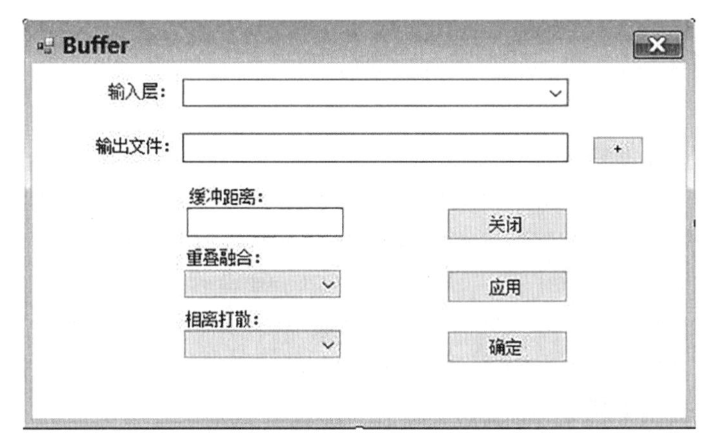


图 15-1　Buffer 窗体控件布局

#### (2) 设置控件属性

设置相应控件的相关属性，见表 15-1。

表 15-1　**Buffer 窗体控件命名表**

| 控件 | Name 属性 | 含义 | 备注 |
|---|---|---|---|
| ComboBox | cbxInputLyr | 输入图层 | |
| TextBox | txtOutput | 输出文件 | shp 格式 |
| TextBox | txtDistance | 缓冲距离 | |
| ComboBox | cbxDissolve | 融合选项 | “YES”，“NO” |
| ComboBox | cbxExplode | 分散选项 | “YES”，“NO” |
| Button | btnBrowser | 输出文件浏览 | |
| Button | btnApp | 执行分析 | |
| Button | btnCancel | 关闭 | |
| Button | btnOK | 确定 | |

**(3)添加 BufferFrm 的全局变量**

```
private IMapControl3 _mapControl;
```

**(4)添加 BufferFrm 事件响应函数、功能函数、辅助函数:**

①添加【应用】按钮 Click 事件响应函数;

②添加【确定】按钮 Click 事件响应函数。

代码如下:

```
public partial class BufferFrm : Form
{
    private IMapControl3 _mapControl = null;
    public BufferFrm(IMapControl3 mapControl)
    {
        InitializeComponent();
        _mapControl = mapControl;
    }
    //事件响应函数
    private void BufferFrm_Load(object sender, EventArgs e)
    private void btnCancel_Click(object sender, EventArgs e)
    private void btnOK_Click(object sender, EventArgs e)
    private void btnBrowser_Click(object sender, EventArgs e)
    private void btnApp_Click(object sender, EventArgs e)
    //核心功能函数
    public IGeometryCollection GetBufferCollection(IFeatureClass
inputFCls, double radius, bool bDissolve, bool bExplode)
    //辅助函数
    private ILayer GetLayerByname(string lyrName)
}
```

### 2. BufferFrm 类的实现

**(1)Load 事件响应函数的实现**

Load 事件响应函数的主要作用是:用 MapControl 中图层名填充 cbxInputLyr(输入图层)控件,代码如下:

```
private void BufferFrm_Load(object sender, EventArgs e)
{
    for (int i = 0; i < _mapControl.Map.LayerCount; i++)
    {
        ILayer aLayer = _mapControl.Map.get_Layer(i);
        if (aLayer is IFeatureLayer)
        {
```

```
            this.cbxInputLyr.Items.Add(aLayer.Name);
        }
    }
    this.cbxDissolve.SelectedIndex = 0;
    this.cbxExplode.SelectedIndex = 0;
}
```

**(2) btnBrowser 按钮响应函数的实现**

实现代码如下：

```
private void btnBrowser_Click(object sender, EventArgs e)
{
    SaveFileDialog openDlg = new SaveFileDialog();
    openDlg.Filter = "Shapefile (*.shp)|*.shp";
    if (openDlg.ShowDialog() == DialogResult.OK)
    {
        string fullFileName = openDlg.FileName;
        this.txtOutput.Text = fullFileName;
    }
}
```

**(3) btnApp 按钮响应函数的实现**

btnApp 按钮响应函数可实现缓冲区分析，步骤如下：

第一步，依据分析参数，获取缓冲区集合，要利用核心函数 GetBufferCollection()；

第二步，创建缓冲区要素类；要利用 GeodatabaseOper 类的功能，参见实验《创建要素类》；

第三步，将缓冲多边形存入输出要素类，如成功则会显示结果。实现代码如下：

```
private void btnApp_Click(object sender, EventArgs e)
{
    //1:获取缓冲区集合
        IFeatureLayer inputFLyr = GetLayerByname(cbxInputLyr.Text)
as IFeatureLayer;
        bool bDissolve = (this.cbxDissolve.Text == "YES") ? true :
false;
        bool bExplode = (this.cbxExplode.Text == "YES") ? true :
false;
        double dDis = double.Parse(this.txtDistance.Text);
        IGeometryCollection outputBuffers = GetBufferCollection
(inputFLyr.FeatureClass, dDis, bDissolve, bExplode);

    //2:创建缓冲区要素
```

```
        //==准备输出目录和要素类名称
        string outputFileName = this.txtOutput.Text;
        string path = System.IO.Path.GetDirectoryName(outputFileName);
        string shpName = System.IO.Path.GetFileName(outputFileName);
        //==使用 GeodatabaseOper 创建输出要素类:
        GeodatabaseOper Oper = new GeodatabaseOper();
        IFeatureWorkspace pWks = Oper.OpenShapefileWorkspace(path)
as IFeatureWorkspace;
        IFeatureClass outputFeatCls = Oper.CreateFeatureClass(null,
pWks, shpName, esriGeometryType.esriGeometryPolygon);

        //3:将缓冲多边形保存到输出要素集
        if (SaveToFeatureClass(outputBuffers, outputFeatCls))
        {
            //添加到图层显示
            IFeatureLayer newFlyr = new FeatureLayerClass();
            newFlyr.FeatureClass = outputFeatCls;
            newFlyr.Name = outputFeatCls.AliasName;
            _mapControl.AddLayer(newFlyr, 0);
            _mapControl.ActiveView.Refresh();
        }
    }

    private void btnCancel_Click(object sender, EventArgs e)
    {
        this.Close();
    }

    private void btnOK_Click(object sender, EventArgs e)
    {
        this.Close();
    }
```

**(4)核心功能函数 GetBufferCollection()的实现**

核心功能函数 GetBufferCollection()可实现输入要素类的缓冲区分析，结果返回多边形集合，基本思路是：

第一步：先用 GeometryBagClass 对输入要素类的几何对象装包。

第二步：构建缓冲区分析工具，包括初始化 IBufferConstruction 接口变量及利用 IBufferConstructionProperties 接口设置分析参数；这里主要设置 UnionOverlappingBuffers（是

否联合重叠多边形，即融合）、ExplodeBuffers（相离多边形是否分散）。

第三步：调用 ConstructBuffers 函数执行缓冲区分析。

代码如下：

```
public IGeometryCollection GetBufferCollection ( IFeatureClass
inputFCls, double
                        radius, bool bDissolve, bool bExplode)
{
    //1:将缓冲对象装包
    IGeometryCollection geometryBag = new GeometryBagClass();
    IFeature feature = null;
    IFeatureCursor fcur = inputFCls.Search(null, false);
    while ((feature = fcur.NextFeature()) ! = null)
    {
        geometryBag.AddGeometry(feature.Shape);
        //调用 Search 函数时,如果回收参数:Recycling = true,必须用 ShapeCopy:
        //geometryBag.AddGeometry(feature.ShapeCopy);
    }

    //2:构建缓冲分析工具
    IBufferConstruction bufferContruction = new Buffer
ConstructionClass();
    IBufferConstructionProperties bufferProper=bufferContruction
as IBufferConstructionProperties;
    bufferProper.EndOption = esriBufferConstructionEndEnum.
esriBufferRound; //圆角
    bufferProper.UnionOverlappingBuffers = bDissolve;//Disslove
    bufferProper.ExplodeBuffers = bExplode;        //Explode

    //3:执行缓冲区分析
    IGeometryCollection outputBuffers = new GeometryBagClass();
    bufferContruction.ConstructBuffers(geometryBag as IEnum
Geometry, radius, outputBuffers);

    return outputBuffers;
}
```

**(5)辅助函数**

SaveToFeatureClass()函数使用 Insert 游标完成，代码如下：

```
public bool SaveToFeatureClass (IGeometryCollection outputBuffers,
```

```
                                IFeatureClass outputFeatCls)
    {
        try
        {
            //获取输出要素类的插入游标
            IFeatureCursor pFCursor = outputFeatCls.Insert(true);

            //遍历 Geometry 集合,每个 Geometry 创建一个要素
            IEnumGeometry enumGeometry = outputBuffers as IEnum
Geometry;
            IGeometry pArea = enumGeometry.Next();
            while (pArea ! = null)
            {
                //创建缓存要素
                IFeatureBuffer pBufferFeature = outputFeatCls.
CreateFeatureBuffer();
                pBufferFeature.Shape = pArea as IPolygon;
                //插入输出要素类
                pFCursor.InsertFeature(pBufferFeature);
                //下一个 Geometry
                pArea = enumGeometry.Next();
            }
        }
        catch (Exception ex)
        {
            MessageBox.Show( ex.Message.ToString());
            return false;
        }
        return true;
    }

    private ILayer GetLayerByname(string lyrName)
    {
        ILayer pLayer = null;
        for (int i = 0; i < _mapControl.LayerCount; i++)
        {
            ILayer tempLayer = _mapControl.get_Layer(i);
            if (tempLayer.Name = = lyrName)
```

```
            {
                pLayer = tempLayer;
                break;
            }
        }

        return pLayer;
    }
```

### 3. 功能调用

在主菜单 Geoprocessing 上添加一菜单项，并命名为 Buffer，创建并修改 Click 事件响应函数，代码如下：

```
    private void bufferToolStripMenuItem1 _ Click ( object sender,
EventArgs e)
    {
        BufferFrm frm = new BufferFrm(m_mapControl);
        if (frm.ShowDialog() = = DialogResult.OK)
        {
            this.m_mapControl.ActiveView.Refresh();
        }
    }
```

## 五、功能测试

①单击 F5 键，编译运行程序；
②加载数据：… \\ Data \\ 矢量数据 \\ 房屋道路 \\ RC. shp；
③点击菜单【Buffer】，启动缓冲分析对话框，设置输出文件名，选择：
◆ 输入图层=RC；
◆ 缓冲距离=100；
◆ 融合参数=ALL；
◆ 打散参数=true。
④点击【应用】按钮，确定后即可看到结果。最终效果如图 15-2 所示。

## 六、思考与练习

①使用 FeatureClass 的插入游标相比使用函数 Store( )保存要素，有何优点？
②GeometryBagClass 类与 PolygonClass 类都可以装 Geometry 对象，有何区别？

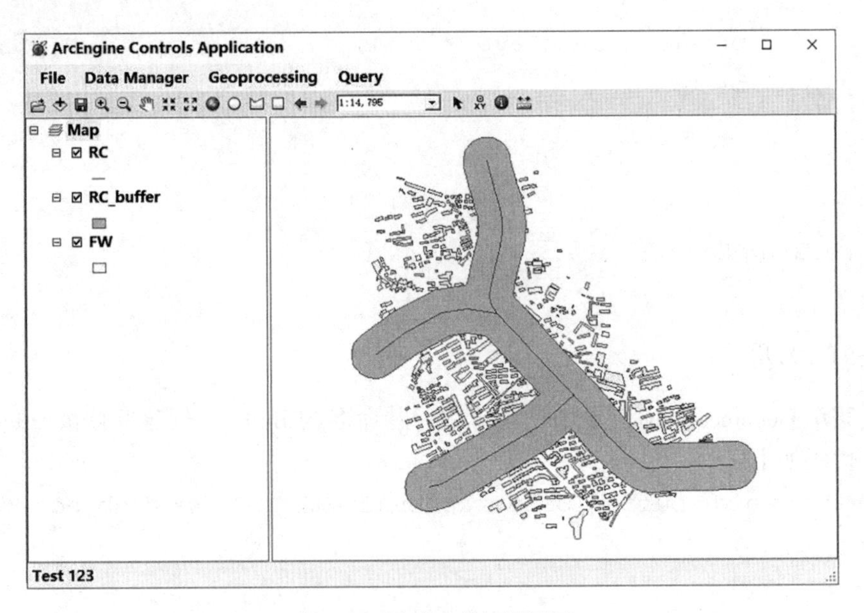


图 15-2　缓冲区分析效果图

# 实验十六　矢量数据叠置分析

## 一、目的与要求

①掌握使用 IBasicGeoprocessor 接口进行叠置分析的步骤和方法；
②加深对 IFeatureClassName 接口的理解。

## 二、实验原理

叠置分析在 ArcEngine 中底层实现属于拓扑运算的范畴，包括裁切(Clip)、求差(Difference)、交集(Intersect)、对称差分(又称为异或，SymmetricDifference)和并集(Union)等，这些拓扑运算在 ArcEngine 的 ITopologicalOperator 接口中定义，高级几何对象(构成要素的几何对象：Multipoint、Polygon 和 Polyline 等)可实现这个接口。

ITopologicalOperator 接口是面向单个几何对象的。ArcEngine 还提供了面向要素类的叠置分析组件 BasicGeoprocessorClass，它实现了 IBasicGeoprocessor 接口，包括 Clip、Dissolve、Intersect、Union、Merge 等方法。自 ArcGIS 9.3 后，ArcEngine 都提供了 GP 工具分析方法，其中 Geoprocessing 组件提供了更加强大的面向要素类的叠置分析工具，但要求具有较高的使用权限。

## 三、实验环境

①开发环境：Visual Studio 2015 + ArcGIS Engine 10.5；
②开发语言：C#；
③实验数据：... \\ Data \\ 矢量数据 \\ Ex4.mdb：
- ◆ 云南县界；
- ◆ Clip_Poly；
- ◆ Grid_Poly；

……

## 四、内容与步骤

本实验可实现叠加分析功能：用户右击【Data Manager】主菜单上【Overlay】菜单项，激活叠加分析对话框，用户可选择输入图层、输出文件，以及控制参数表等。

### 1. OverlayAnalyst 窗体设计

#### (1)添加叠置分析对话框类

新建一个 Windows 窗体，并命名为“OverlayAnalysisFrm”，修改窗体的 Text 属性为“OverlayAnalysisFrm”，并添加 Button、Label、ComboBox 等控件。控件布局如图 16-1 所示。

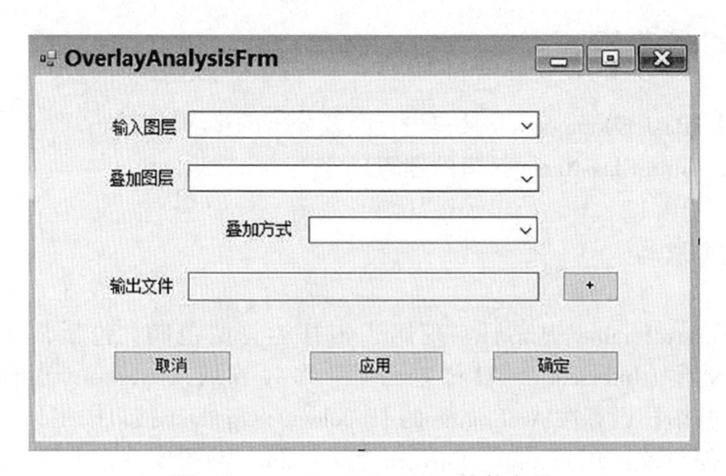


图 16-1　OverlayAnalysisFrm 控件布局

#### (2)设置控件属性

设置相应控件的相关属性，见表 16-1。

表 16-1　**OverlayAnalysisFrm 控件命名表**

| 控件 | Name | 含义 | 备注 |
| --- | --- | --- | --- |
| ComboBox | cbxInputLayers | 输入图层 | |
| ComboBox | cbxOverlayLayers | 叠置图层 | |
| ComboBox | cbxOverlayMethod | 叠置方法 | |
| TextBox | txtOutput | 输出文件名 | Shp 格式 |
| Button | btnExplor | 文件浏览 | |
| Button | btnApp | 应用 | |
| Button | btnCancel | 关闭 | |
| Button | btnOK | 确定 | |

#### (3)添加 OverlayAnalysisFrm 的全局变量

```
private IMapControl3 _mapControl;
```

**(4)添加 OverlayAnalysisFrm 事件响应函数、功能函数、辅助函数**

①添加应用按钮 Click 事件响应函数;

②添加确定按钮 Click 事件响应函数。

代码如下:

```
public partial class OverlayAnalysisFrm : Form
{
    private IMapControl3 _mapControl = null;
    public OverlayAnalysisFrm(IMapControl3 mapControl)
    {
        InitializeComponent();
        _mapControl = mapControl;
    }

    //事件响应函数
    private void OverlayAnalysisFrm_Load(object sender, EventArgs e)
    private void btnExplore_Click(object sender, EventArgs e)
    private void btnApp_Click(object sender, EventArgs e)
    private void btnOK_Click(object sender, EventArgs e)
    private void btnCancel_Click(object sender, EventArgs e)

    //功能函数
    private IFeatureClass Intersect(IFeatureLayer inputfeatureLyr,
IFeatureLayer overlayfeatureLyr,IFeatureClassName outputFeature
ClassName)
    public IFeatureClass Clip (IFeatureLayer inputfeatureLyr,
IFeatureLayer clipfeatureLyr,IFeatureClassName outputFeature Class
Name)
    public IFeatureClass Union (IFeatureLayer inputfeatureLyr,
IFeatureLayer overlayfeatureLyr, IFeatureClassName outputFeature
ClassName)

    //辅助函数
    private IFeatureLayer GetFeatureLayer(string layerName)
    private IEnumLayer GetFeatureLayers()
    private IFeatureClassName CreateFeatureClassName(stringpath
Name, string dsName, esriGeometryType geoType)

}
```

2. OverlayAnalysisFrm 类的实现

**(1)Load 事件响应函数的实现**

Load 事件响应函数的主要作用是：用 MapControl 中矢量图层名填充 cbxInputLayers、cbxOverlayLayers(图层名)控件，("Intersect","Clip","Union")填充 cbxOverlayMethod 控件。代码如下：

```
private void OverlayAnalysisFrm_Load(object sender, EventArgs e)
{
    IEnumLayer pLayers = GetFeatureLayers();
    ILayer pLyr = null;
    while ((pLyr = pLayers.Next()) ! = null)
    {
        cbxInputLayers.Items.Add(pLyr.Name);
        cbxOverlayLayers.Items.Add(pLyr.Name);
    }

    this.cbxOverlayMethod.Items.Add("Intersect");
    this.cbxOverlayMethod.Items.Add("Clip");
    this.cbxOverlayMethod.Items.Add("Union");
    this.cbxOverlayMethod.SelectedIndex = 0;
}
```

**(2)btnExplore 按钮响应函数的实现**

btnExplore 按钮响应函数的实现代码如下：

```
private void btnExplore_Click(object sender, EventArgs e)
{
    SaveFileDialog saveDlg = new SaveFileDialog();
    {
        saveDlg.CheckPathExists = true;
        saveDlg.Filter = "Shapefile (*.shp)|*.shp";
        saveDlg.OverwritePrompt = true;
        saveDlg.Title = "Output Layer";
        saveDlg.RestoreDirectory = true;
        saveDlg.FileName = cbxInputLayers.Text +"_"+ cbxOverlay
Method.Text + ".shp";
    }

    if (saveDlg.ShowDialog() = = DialogResult.OK)
    {
```

```
            txtOutput.Text = saveDlg.FileName;
        }
    }
```

**(3) btnApp 按钮响应函数的实现**

btnApp 按钮响应函数完成叠加分析操作，步骤如下：

第一步，从界面获取分析参数，包括分析方法、输入图层、叠加图层、输出文件名；

第二步，创建要素类名称对象，结果将存储到它所代表的要素类中；

第三步，调用分析方法相应的功能函数：Intersect( )或 Union( )或 Clip( )；

第四步，添加到地图。

代码如下：

```
    private void btnApp_Click(object sender, EventArgs e)
    {
        //获取参数
        string overlayMethod = this.cbxOverlayMethod.Text;
        IFeatureLayer inputfeatureLyr = GetFeatureLayer(cbxInput
Layers.Text);
        IFeatureLayer overlayfeatureLyr = GetFeatureLayer(cbxOverlay
Layers.Text);

        string outputFileName = this.txtOutput.Text;
        string pathName = System.IO.Path.GetDirectoryName(output
FileName);
        string shpName = System.IO.Path.GetFileName(outputFileName);

        //创建要素类名称对象
        esriGeometryType type = inputfeatureLyr.Feature
Class.ShapeType;
        IFeatureClassName outputFeatureClassName = CreateFeature
ClassName(pathName, shpName, type);

        //执行叠加分析
        IFeatureClass result = null;
        switch (overlayMethod)
        {
            case "Intersect":
                result = Intersect(inputfeatureLyr, overlayfeature
Lyr, outputFeatureClassName);
                break;
```

```
            case "Union":
                result = Union(inputfeatureLyr, overlayfeatureLyr,
outputFeatureClassName);
                break;
            case "Clip":
                result = Clip(inputfeatureLyr, overlayfeatureLyr,
outputFeatureClassName);
                break;
            default:
                break; ;
        }

        //添加到地图
        IFeatureLayer newFlyr = new FeatureLayerClass();
        newFlyr.FeatureClass = result;
        newFlyr.Name = result.AliasName;
        _mapControl.AddLayer(newFlyr, 0);
        _mapControl.ActiveView.Refresh();
    }

    private void btnOK_Click(object sender, EventArgs e)
    {


    }
    private void btnCancel_Click(object sender, EventArgs e)
    {
        this.Close();
    }
```

**(4)功能函数**

Intersect( )、Clip ( )、Union ( ) 功能函数实现方法类似，步骤是：先初始化 IBasicGeoprocessor 接口(该接口有个 SpatialReference 属性，代表分析结果的空间参考系)，然后调用 IBasicGeoprocessor 对应方法即可。代码如下：

```
    private IFeatureClass Intersect(IFeatureLayer inputfeatureLyr,
IFeatureLayer overlayfeatureLyr, IFeatureClassName outputFeature
ClassName)
    {
        try
        {
```

```
            double tol = 0.001; //几何处理容差
            IBasicGeoprocessor basicGeoprocessor = new Basic
GeoprocessorClass();
            IGeoDataset gDataset = inputfeatureLyr.FeatureClass as
IGeoDataset;
            basicGeoprocessor.SpatialReference = gDataset.Spatial
Reference;

            //得到相交的几何图形
            IFeatureClass outFeatureClass = basicGeoprocessor.
Intersect(inputfeatureLyr as ITable, false, overlayfeatureLyr as
ITable, false, tol, outputFeatureClassName);
            return outFeatureClass;
        }
        catch (Exception ms)
        {
            throw ms;
        }
    }

    public IFeatureClass Clip(IFeatureLayer inputfeatureLyr, IFeature
Layer clipfeatureLyr,IFeatureClassName outputFeatureClassName)
    {
        try
        {
            double tol = 0.001;
            IBasicGeoprocessor basicGeoprocessor = new Basic
GeoprocessorClass();
            IGeoDataset gDataset = inputfeatureLyr.FeatureClass as
IGeoDataset;
            basicGeoprocessor.SpatialReference = gDataset.Spatial
Reference;

            //裁剪得到重叠的几何图形
             IFeatureClass outFeatureClass = basicGeoprocessor.Clip
(inputfeatureLyr as ITable, false, clipfeatureLyr as ITable, false,
tol, outputFeatureClassName);
            return outFeatureClass;
```

```
        }
        catch (Exception ms)
        {
            throw ms;
        }
    }

    public IFeatureClass Union (IFeatureLayer inputfeatureLyr,
IFeatureLayer overlayfeatureLyr, IFeatureClassName outputFeature
ClassName)
    {
        try
        {
            double tol = 0.001;
            IBasicGeoprocessor basicGeoprocessor = new Basic
GeoprocessorClass();
            IGeoDataset gDataset = inputfeatureLyr.FeatureClass as
IGeoDataset;
            basicGeoprocessor.SpatialReference = gDataset.Spatial
Reference;

            //得到重叠的几何图形
            IFeatureClass outputFeatureClass = basicGeoprocessor.Union
(inputfeatureLyr as ITable, false, overlayfeatureLyr as ITable,
false, tol, outputFeatureClassName);
            return outputFeatureClass;
        }
        catch (Exception ms)
        {
            throw ms;
        }
    }
```

**(5)辅助函数**

函数 CreateFeatureClassName()可创建一个要素类的名称对象(IFeatureClassName), IName 对象是一个代表性对象。通过使用 IName 对象,可以访问它所代表的对象的一些基本属性,而不用将整个对象调入内存,这样可以节省资源,提高程序运行效率。

代码如下:

```
    private IFeatureClassName CreateFeatureClassName(string pathName,
```

```
string dsName, esriGeometryType geoType)
    {
        IFeatureClassName ftClassName = new FeatureClassNameClass();
        {
            ftClassName.FeatureType = esriFeatureType.esriFTSimple;
            ftClassName.ShapeFieldName = "Shape";
            ftClassName.ShapeType = geoType;
        }

        //工作空间名称
        IWorkspaceName pNewWSName = new WorkspaceNameClass();
        {
            pNewWSName.WorkspaceFactoryProgID = "esriDataSourcesFile.
ShapefileWorkspaceFactory";
            pNewWSName.PathName = pathName;
        }

        //数据集名称
        IDatasetName pDatasetName = ftClassName as IDatasetName;
        {
            pDatasetName.Name = dsName;
            pDatasetName.WorkspaceName = pNewWSName;
        }

        return ftClassName;
    }
```

GetFeatureLayers()函数可获取地图所有矢量数据图层集合，GetFeatureLayer()根据图名获取对应图层的接口，代码如下：

```
    private IFeatureLayer GetFeatureLayer(string layerName)
    {
        //get the layers from the maps
        IEnumLayer layers = GetFeatureLayers();
        layers.Reset();

        ILayer layer = null;
        while ((layer = layers.Next()) ! = null)
        {
```

```
            if (layer.Name == layerName)
                return layer as IFeatureLayer;
        }

        return null;
    }

    //获取所有要素层的枚举器
    private IEnumLayer GetFeatureLayers()
    {
        UID uid = new UIDClass();
        uid.Value = "{40A9E885-5533-11d0-98BE-00805F7CED21}";
        IEnumLayer layers = _mapControl.Map.get_Layers(uid, true);

        return layers;
    }
```

### 3. 功能调用

在主菜单【Geoprocessing】上添加一菜单项(命名为"Overlay")，创建并修改 Click 事件响应函数，代码如下：

```
private void overlayToolStripMenuItem_Click(object sender, EventArgs e)
{
    OverlayAnalystFrm frm = new OverlayAnalystFrm(m_mapControl);
    if (frm.ShowDialog() == DialogResult.OK)
    {
        this.m_mapControl.ActiveView.Refresh();
    }
}
```

## 五、编译测试

①单击 F5 键，编译运行程序。

②加载数据：… \\ Data \\ 矢量数据 \\ Ex4. mdb \\ Ex4. mdb \\ ：

- 云南县界；
- Clip_Poly。

③点击菜单【Overlay】启动叠加分析对话框，设置输出文件名，选择：

- 输入图层= 云南县界；

◆　叠加图层= Clip_Poly;

◆　分析方法= Clip。

④点击【应用】按钮，即可看见叠加结果已添加到地图。Clip 分析效果图如图 16-2 所示。

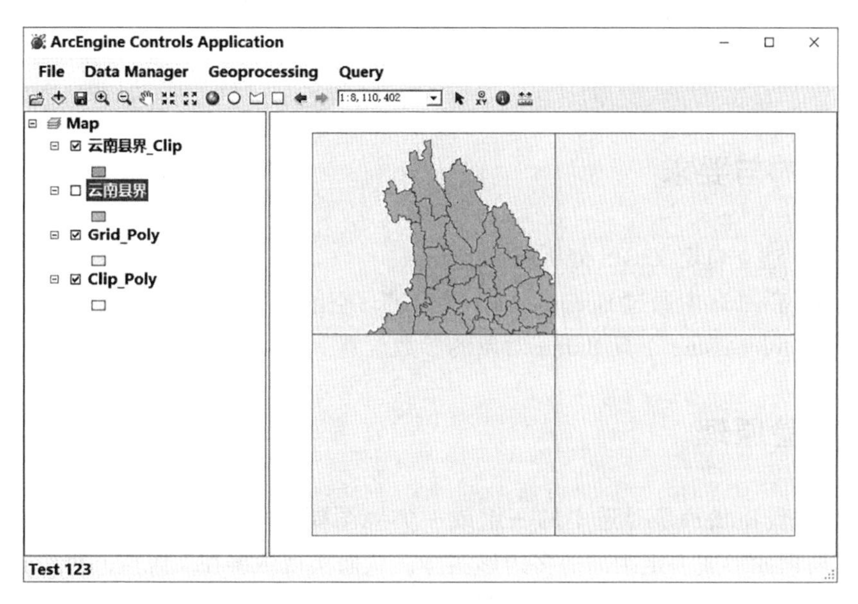


图 16-2　Clip 分析效果图

## 六、扩展练习

①试利用 IBasicGeoprocessor 实现图层合并功能。

②试利用 IBasicGeoprocessor 接口实现求差功能。

# 实验十七　缓冲区分析(GP)

## 一、目的与要求

①掌握 C# 建立自定义 GIS 分析组件的方法；
②熟悉 ArcEngine 中通过 Geoprocessor 接口进行空间分析的基本步骤；
③掌握 Geoprocessing 工具 Buffer 对象的参数配置。

## 二、实验原理

缓冲区分析(Buffer)是对选中的一组或一类地图要素(点、线或面)按设定的距离条件，围绕其要素而形成一定距离的多边形实体，从而实现数据在二维空间得以扩展的信息分析方法。缓冲区应用的实例有：污染源对其周围的污染量随距离而减小，确定污染的区域；为失火建筑找到距其 500m 范围内所有的消防水管等。

ArcEngine 9.2 之后支持通过代码完成既有工具箱中工具的调用(至于需要添加哪几个 Reference 以及 using 命名空间请查看帮助文档)，步骤如下：

①创建一个 Geoprocessor 的类对象 GP，由它来执行 Geoprocessing 的工具；
②创建一个 Geoprocessing 工具的类对象，比如这里的 Buffer 工具的类对象 Buffer；
③填写一些参数，参数分为 in 和 out，Required 和 Optional；
④调用 GP 的 Execute 方法执行。

## 三、实验环境

①开发环境：Visual Studio 2015 + ArcGIS Engine 10.5。
②开发语言：C#。
③实验数据：… \\ Data \\ 矢量数据 \\ 房屋道路：

- FW. shp；
- RC. shp；

……

# 四、内容与步骤

## 1. 建立“功能扩展”组件

VS 2015 执行菜单【new】→【Project】，在弹出如图 17-1 的对话框中选择 C#类库，命名为“BufferAnalyst”。

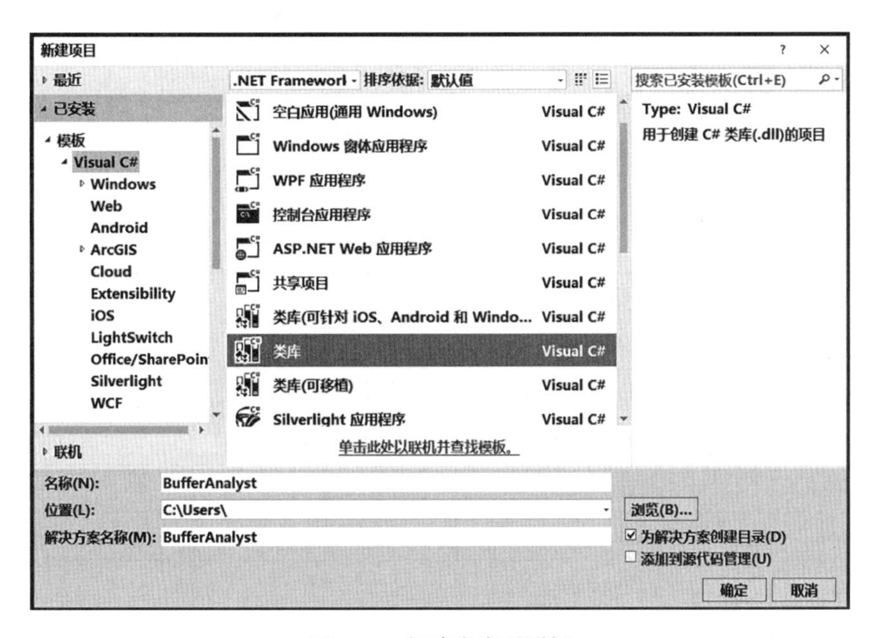


图 17-1　新建类库对话框

## 2. 添加缓冲分析窗口类

### (1)新建 Windows From 类

新建 Windows From 类，并命名为“BufferDlg”，如图 17-2 所示。

### (2)添加界面元素

设置相应控件的相关属性，见表 17-1。

### (3)添加功能函数

①添加私有成员 IMapControl 3m、mapControl;

②构造函数添加参数 IMapControl 3mapControl，用于赋值 mapControl;

③添加 Load 事件函数，输出按钮 Click 事件函数，分析执行按钮 Click 事件函数等。

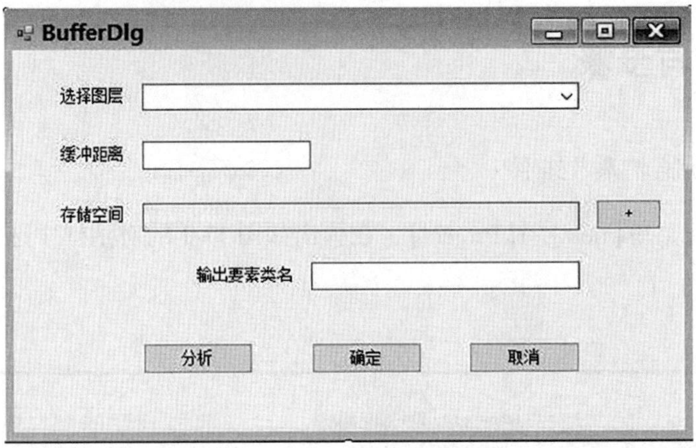

图 17-2 BufferDlgp 窗体控件布局

表 17-1 **BufferDlg 窗体控件命名表**

| 控件 | Name 属性 | 含义 | 备注 |
|---|---|---|---|
| ComboBox | cbxLayers | 图层名 | |
| TextBox | txtBufferDistance | 缓冲距离 | |
| TextBox | txtOutputPath | 输出路径 | ReadOnly |
| TextBox | TxtFCname | 输出要素类名称 | |
| Button | btnOutputPath | 输出路径设置 | |
| Button | btnBuffer | 分析 | |
| Button | btnCancel | 关闭 | |

类设计如下：

```
public partial class BufferDlg_ : Form
{
    private IMapControl3 m_mapControl = null;
    public BufferDlg_(IMapControl3 mapControl)
    {
        InitializeComponent();
        m_mapControl = mapControl;

        IAoInitialize pao = new AoInitializeClass();
        pao.Initialize(esriLicenseProductCode.esriLicenseProductCodeAdvanced);
    }
```

```
        //Load 事件响应函数
        private void BufferDlg_Load(object sender, EventArgs e)
        //图层选择响应函数
        private void cbxLayers_SelectedIndexChanged(object sender,
EventArgs e)

        //输出设置按钮 Click 事件响应函数
        private void btnOutputPath_Click(object sender, EventArgs e)
        //分析执行按钮 Click 事件响应函数
        private void btnBuffer_Click(object sender, EventArgs e)

        //确定、取消按钮 Click 事件响应函数
        private void btnCancel_Click(object sender, EventArgs e)
        private void btnOK_Click(object sender, EventArgs e)
    }
```

### 3. 功能实现

**(1) Load 事件响应函数**

Load 事件响应函数代码如下:

```
private void BufferDlg_Load(object sender, EventArgs e)
{
    //load all the feature layers in the map to the layers combo
    IEnumLayer layers = GetLayers();
    layers.Reset();
    ILayer layer = null;
    while ((layer = layers.Next()) != null)
    {
        cbxLayers.Items.Add(layer.Name);
    }
    //select the first layer
    if (cbxLayers.Items.Count > 0)
        cbxLayers.SelectedIndex = 0;

    string tempDir = System.IO.Path.GetTempPath();
    txtOutputPath.Text = System.IO.Path.Combine(tempDir, "");
}

private void cbxLayers_SelectedIndexChanged(object sender,
```

```
EventArgs e)
    {
        txtFCname.Text = cbxLayers.SelectedItem.ToString() + "_
buffer";
    }

    private void btnOK_Click(object sender, EventArgs e)
    {
        this.Close();
    }

    private void btnCancel_Click(object sender, EventArgs e)
    {
        this.Close();
    }
```

**(2) btnBuffer_Click 函数实现**

ArcEngine 9.2 之后支持通过代码完成既有工具箱中工具的调用(至于需要添加哪几个 Reference 以及 using 命名空间请参看帮助文档)，步骤如下：

第一步，构建一个 Geoprocessor 的类对象 GP，用于执行 Geoprocessing 的工具。

第二步，构建一个 Geoprocessing 工具 Buffer 类对象。

第三步，工具参数赋值，参数如下：

- in_features：输入要素集，可赋值图层；
- out_feature_Class：输出要素类，可赋值全路径 shp 文件名。如果仅赋值要素类名，需要使用 Geoprocessor 的 SetEnvironmentValue("Key", "Value") 函数，设置"工作空间"(按路径表示：如 shp 存放目录，gdb 数据库名)；
- buffer_distance_or_field：缓冲距离，可指定距离，或由属性字段决定；
- dissolve_option：融合参数，一般选择"All"。

第四步，调用 GP 的 Execute 方法执行工具。

代码如下：

```
private void btnBuffer_Click(object sender, EventArgs e)
{
    ILayer selectedLyr = GetFeatureLayer(cbxLayers.SelectedItem.
ToString());
    if(selectedLyr == null)
    {
        MessageBox.Show("Bad Layer!");
        return;
    }
```

```
        //转换 distance 为 double 类型
        double bufferDistance = 0;
        double.TryParse(txtBufferDistance.Text, out bufferDistance);
        if (0.0 == bufferDistance)
        {
            MessageBox.Show("Bad buffer distance!");
            return;
        }

        //step1: get an instance of the geoprocessor
        Geoprocessor gp = new Geoprocessor();
        gp.OverwriteOutput = true;
        gp.SetEnvironmentValue("workspace", this.txtOutputPath.
Text);

        //step2: create a new instance of a buffer tool
        ESRI.ArcGIS.AnalysisTools.Buffer buffer = null;
        buffer = new ESRI.ArcGIS.AnalysisTools.Buffer();

        //step3: set parameter of tool
        buffer.in_features = selectedLyr;
        buffer.out_feature_class = this.txtFCname.Text;
        string strUnite = esriUnits.esriMeters.ToString().Substring
(4);
        buffer.buffer_distance_or_field = bufferDistance.ToString
() + " " + strUnite;
        buffer.dissolve_option = "ALL";
        //buffer.line_side = "FULL";
        //buffer.line_end_type = "ROUND";

        //step4: execute the buffer tool
        IGeoProcessorResult results = null;
        string strErrorInfo = " Finished to buffer layer: " +
selectedLyr.Name + "\r\n";
        try
        {
            results = (IGeoProcessorResult)gp.Execute(buffer, null);
```

```
        }
        catch (Exception ex)
        {
            strErrorInfo = "Failed to buffer layer: " + selectedLyr.
Name + "\r\n";
        }
        MessageBox.Show(strErrorInfo);
    }
```

**(3) btnOutputPath_Click 函数实现**

实现代码如下：

```
    private void btnOutputPath_Click(object sender, EventArgs e)
    {
        //FolderBrowserDialog 支持 gdb 数据库目录、shapefile 存储目录
        //如果是 mdb 数据库,用文件浏览对话框 OpenFileDialog
        FolderBrowserDialog openFolder = new FolderBrowserDialog();
        DialogResult dr = openFolder.ShowDialog();
        if (dr == DialogResult.OK)
            txtOutputPath.Text = openFolder.SelectedPath;
    }
```

**(4) 辅助函数**

辅助函数实现代码如下：

```
    private IFeatureLayer GetFeatureLayer(string layerName)
    {
        //get the layers from the maps
        IEnumLayer layers = GetLayers();
        layers.Reset();

        ILayer layer = null;
        while ((layer = layers.Next()) != null)
        {
            if (layer.Name == layerName)
                return layer as IFeatureLayer;
        }

        return null;
    }

    private IEnumLayer GetLayers()
```

```
{
    UID uid = new UIDClass();
    uid.Value = "{40A9E885-5533-11d0-98BE-00805F7CED21}";
    IEnumLayer layers = m_mapControl.Map.get_Layers(uid, true);

    return layers;
}
```

3. 集成

①在主程序中添加“BufferAnalyst”组件的引用；

②在【Geoprocessing】主菜单上添加【Buffer(GP)】菜单项，并创建 Click 响应函数：bufferAnalystToolStripMenuItem_Click(…)

③在响应函数中添加如下代码：

```
private void bufferAnalystToolStripMenuItem_Click(object sender,
EventArgs e)
{
    BufferDlg frm = new BufferDlg (m_mapControl);
    frm.ShowDialog();
}
```

## 五、功能测试

①单击 F5 键，编译运行程序；

②加载数据：… \\ Data \\ 矢量数据 \\ 房屋道路 \\ RC. shp；

③点击菜单“Buffer(GP)”，启动缓冲分析对话框，设置输出路径、输出文件名，选择：

- ◆ 输入图层 = RC；
- ◆ 缓冲距离 = 100；

④点击【分析】按钮，当出现“分析完成”提示时，表示工作完成；

⑤添加新的图层。

最终效果如图 17-3 所示。

## 六、思考与练习

①怎样为 Buffer 工具对象按属性字段指定缓冲距离？

②Geoprocessor 的 Excute()函数返回的 IGeoProcessorResult 值有何意义？

③如果输入数据和输出数据都需要放在统一的 gdb 数据库中，该如何设置？

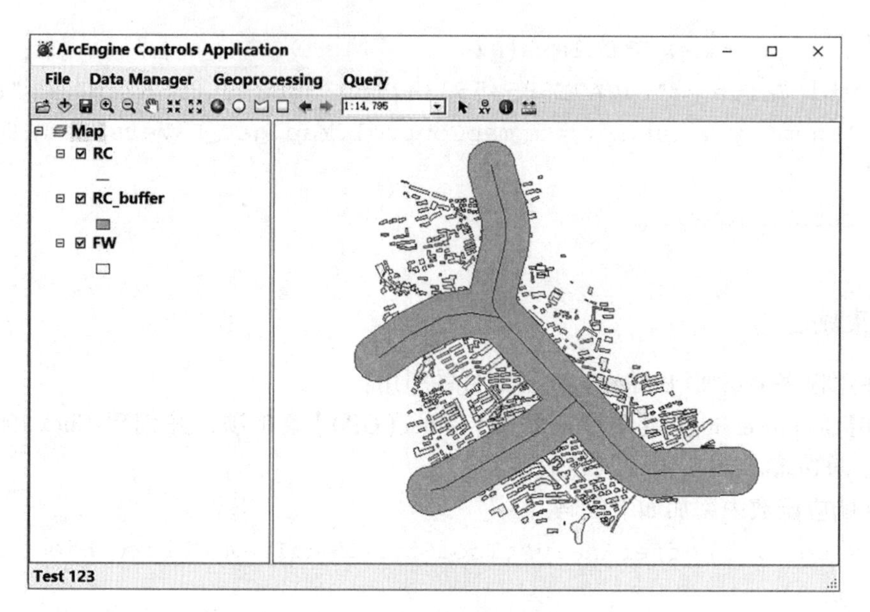


图 17-3　缓冲区分析效果图

# 实验十八　矢量数据叠置分析(GP)

## 一、目的与要求

①掌握使用 Geoprocessor 代理类运行 Geoprocessing 工具进行叠加分析的步骤；
②加深对双输入类型工具+多输入类型工具输入参数的理解；
③理解 Geoprocessor 类对分析环境设置要点。

## 二、实验原理

叠置分析是 GIS 中一种常见的分析功能，它是将组成有关主题的各个数据层进行叠置产生一个新的数据层面，其结果综合了原来两个或多个层所具有的信息，叠置分析不仅生成了新的空间关系，而且还将输入的多个数据层的属性联系起来产生了新的属性关系。

ArcEngine 还提供了面向单个几何对象的底层分析接口：ITopologicalOperator；面向数据集的接口：IBasicGeoprocessor，该接口提供了 Clip、Dissolve、Intersect、Union、Merge 等方法。为降低 ArcObjects 开发者的复杂度，从 ArcGIS 9.3 后，ArcEngine 提供了 GP 工具分析方法，其中 Geoprocessing 组件提供了数据分析、数据管理和数据转换等，包括了上百个 Geoprocessing 工具；由 Geoprocessor 对象可以方便地调用 Geoprocessing 中提供的各类工具。它是执行 ArcGIS 中 Geoprocessing 工具的唯一入口。

使用 GP 工具的步骤是：
①构建一个 Geoprocessor 的类对象 GP，将由它来执行 Geoprocessing 的工具。
②构建一个 Geoprocessing 工具的类对象，比如 Intersect 工具的类对象 Intersect。
③为工具填写参数，参数分为 in 和 out，Required 和 Optional。
④调用 GP 的 Execute 方法执行 Geoprocessing 工具。
本章介绍使用 GP 工具进行叠加分析的过程。

## 三、实验环境

①开发环境：Visual Studio 2015 + ArcGIS Engine 10.5。
②开发语言：C#。
③实验数据：... \\ Data \\ 矢量数据 \\ 房屋道路：
◆　FW.shp；

◆ RC_Buffer. shp；

……

## 四、内容与步骤

本实验实现矢量数据叠置分析的功能：用户右键单击【Geoprocessing】主菜单上【Overlay(GP)】菜单项，激活叠加分析对话框，用户可选择输入图层、输出文件，以及控制参数表等。

### 1. OverlayAnalysis 窗体设计

#### (1)添加叠加分析类对话框类

新建一个 Windows 窗体，并命名为"OverlayAnalysisFrm"，修改窗体的 Text 属性为"OverlayAnalysisFrm"，并添加 Button、TextBox、ComboBox 等控件。控件布局如图 18-1 所示。

图 18-1 OverlayAnalysisFrm 控件布局

#### (2)设置控件属性

表 18-1 **OverlayAnalysisFrm 控件命名表**

| 控件类型 | Name 属性 | 含义 | 备注 |
| --- | --- | --- | --- |
| ComboBox | cbxInputLayers | 输入要素 | |
| ComboBox | cbxOverlayLayers | 叠置要素 | |

续表

| 控件类型 | Name 属性 | 含义 | 备注 |
|---|---|---|---|
| TextBox | txtOutputPath | 叠置结果的输出路径 | |
| Button | btnOutputPath | 选择输出路径 | |
| Button | btnAnalist | 进行叠置分析 | |
| Button | btnCancel | 取消 | |
| GroupBox | | | 作为 txtMessage 的容器 |

**(3)添加 OverlayAnalysisFrm 的全局变量**

```
private IMapControl3  m_mapControl = null;
```

**(4)添加 OverlayAnalysisFrm 事件的响应函数、功能函数、辅助函数**

①添加应用按钮 Click 事件响应函数;

②添加确定按钮 Click 事件响应函数;

代码如下:

```
public class OverlayAnalysisFrm
{
    private IMapControl3  m_mapControl = null;
    public OverlayAnalysisFrm(IMapControl3 mapControl)
    {

        InitializeComponent();
        m_mapControl = mapControl;
    }

    //窗体加载时触发事件,执行函数
    private voidOverlayAnalysisFrm_Load(object sender, EventArgs e)
    //点击应用按钮时,执行函数
    private void btnAnalyst_Click(object sender, EventArgs e)
    //点击确定按钮时,执行函数
    private void buttonOK_Click(object sender, EventArgs e)
    //点击输出路径按钮时,执行函数
    private void btnOutputPath_Click(object sender, EventArgs e)

    //若干功能函数
    private GeoprocessingCreateGeoprocessorTool()
    .....
}
```

### 2. OverlayAnalysisFrm 类的实现

**(1)载入响应函数 OverlayAnalysisFrm_Load()**

OverlayAnalysisFrm 在载入时，需要用 Map 图层名填充 cbxInputLayers，cbxOverlayLayers。具体代码如下：

```
private void OverlayAnalysisFrm_Load(object sender, EventArgs e)
{
    //load all the feature layers in the map to the layers combo
    IEnumLayer layers = GetFeatureLayers();
    ILayer layer = null;
    while ((layer = layers.Next()) ! = null)
    {
        cbxInputLayers.Items.Add(layer.Name);
        cbxOverlayLayers.Items.Add(layer.Name);
    }
}
```

**(2)输出路径设置响应函数 btnOutputLayer_Click()**

输出路径设置由 SaveFileDialog 实现，实现代码如下：

```
private void btnOutputPath_Click(object sender, EventArgs e)
{
    //set the output layer
    SaveFileDialog saveDlg = new SaveFileDialog();
    saveDlg.CheckPathExists = true;
    saveDlg.Filter = "Shapefile (*.shp)|*.shp";
    saveDlg.OverwritePrompt = true;
    saveDlg.Title = "Output Layer";
    saveDlg.RestoreDirectory = true;
    saveDlg.FileName = (string)cbxOverlayLayers.SelectedItem + "
_overlay.shp";

    DialogResult dr = saveDlg.ShowDialog();
    if (dr == DialogResult.OK)
        txtOutputPath.Text = saveDlg.FileName;
}
```

**(3)分析响应函数 btnAnalyst_Click()**

分析响应函数负责执行指定类型的叠加分析操作。步骤如下(注意这里概括为三步)：

①创建 Geoprocessor 代理类；

②创建 Geoprocessing 叠加分析工具，由函数 CreateGeoprocessorTool()完成；

③执行叠加分析。

具体代码如下:

```
private void btnAnalyst_Click(object sender, EventArgs e)
{
    //1:Setting up the Geoprocessor
    Geoprocessor GP = new Geoprocessor();
    GP.OverwriteOutput = true;

    //2:create a new instance of a tool
    IGPProcess gpTool = CreateGeoprocessorTool( " ");

    //3:runtool
    IGeoProcessorResult results = null;
    string strErrorInfo = "Finished to buffer layer: " + "\r\n";
    try
    {
        results = (IGeoProcessorResult)gp.Execute(gpTool , null);
    }
    catch (Exception ex)
    {
        strErrorInfo = "Failed to buffer layer: "  + "\r\n";
    }

    MessageBox.Show(strErrorInfo);
}
```

**(4) CreateGeoprocessorTool( )函数**

Identity、Clip、Erase工具是针对两个要素的运算，需要对输入要素in_features和叠加要素(如identity_features)分别赋值。以下代码可创建一个Clip工具，创建其他工具方法与之类似:

```
private IGPProcess CreateGeoprocessorTool(string strOverlay)
{
    IFeatureLayer inputLayer =
            GetFeatureLayer((string)cbxInputLayers.SelectedItem);
    IFeatureLayer overlayLayer =
            GetFeatureLayer((string)cbxOverlayLayers.SelectedItem);
    string strOutputPath = this.txtOutputPath.Text.ToString();//
输出文件名
```

```
        //创建工具
        ESRI.ArcGIS.AnalysisTools.Clip clipTool = new Clip();
        //设置工具参数
        clipTool.in_features = inputLayer;
        clipTool.clip_features = overlayLayer;
        clipTool.out_feature_class = strOutputPath;
        //clipTool.cluster_tolerance = "0.01";
        //返回
        return (clipTool as IGPProcess);
    }
```

**(5) GetFeatureLayers( )函数**

GetFeatureLayers( )函数获取地图所有矢量数据图层集合，GetFeatureLayer( )函数根据图名获取对应图层的接口，代码如下：

```
    private IFeatureLayer GetFeatureLayer(string layerName)
    {
        //get the layers from the maps
        IEnumLayer layers = GetFeatureLayers();
        layers.Reset();

        ILayer layer = null;
        while ((layer = layers.Next()) ! = null)
        {
            if (layer.Name = = layerName)
                return layer as IFeatureLayer;
        }

        return null;
    }

    //获取所有要素层的枚举器
    private IEnumLayer GetFeatureLayers()
    {
        UID uid = new UIDClass();
        uid.Value = "{40A9E885-5533-11d0-98BE-00805F7CED21}";
        IEnumLayer layers = _mapControl.Map.get_Layers(uid, true);

        return layers;
    }
```

### 3. 功能调用

【Geoprocessing】主菜单上添加 Overlay(GP)菜单项，建立“Overlay”响应函数，具体代码如下：

```
private void overlayToolStripMenuItem _ Click ( object sender, EventArgs e)
{
    OverlayAnalysisFrm frm = new OverlayAnalysisFrm (m_mapControl.Map);
    frm.Show();
}
```

## 五、编译测试

①单击 F5 键，编译运行程序。

②添加数据：… \\ Data \\ 矢量数据 \\ 房屋道路 \\ ：

- ◆ FW. shp；
- ◆ RC_Buffer. shp。

③点击菜单【Overlay(GP)】，启动叠加分析对话框，设置输出文件名，选择：

- ◆ 输入图层= FW；
- ◆ 叠加图层= RC_Buffer；
- ◆ 分析方法= Clip。

④点击【应用】按钮，即可看见叠加结果已添加到地图上。

最终效果如图 18-2 所示。

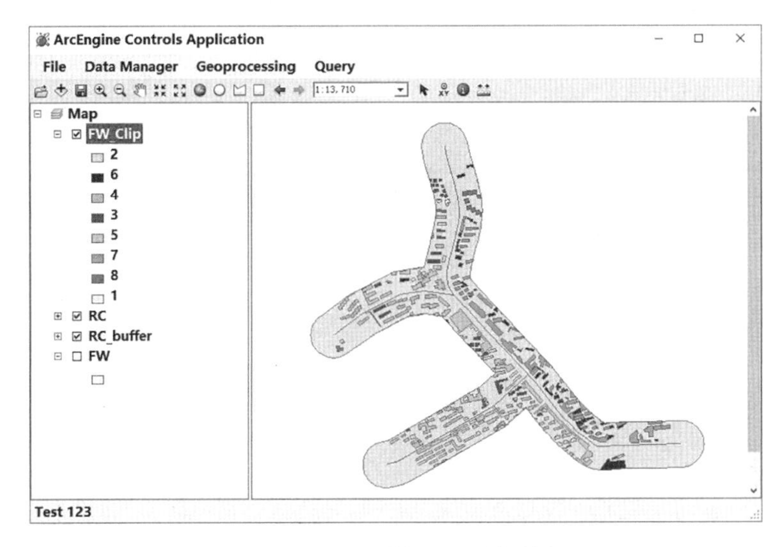


图 18-2　叠置分析(GP)效果图

## 六、思考与练习

①试分别实现 Erase、IDentity、Intersect、Union 分析。
②如果输入数据和输出数据都需要放在统一的 gdb 数据库，该如何设置？

# 实验十九　栅格数据重分类

## 一、目的与要求

①掌握 IReclassOp 接口进行栅格数据重分类的方法；
②熟悉 DataGridView 表格控件的使用方法。

## 二、实验原理

重分类是依据一定规则将原有栅格像元值重新分为若干类别，并将每一分类变换为一个新值并输出。在 ArcGIS Engine 中，RasterReclassOpClass 类实现了栅格数据的重分类。该类实现的两个主要的接口是 IRasterAnalysisEnvironment 接口和 IReclassOp 接口。IReclassOp 接口的 ReclassByRemap 方法是最常用的重分类方法，它依据映射表将旧值变换为新值。

## 三、实验环境

①开发环境：Visual Studio 2015 + ArcGIS Engine 10.5；
②开发语言：C#；
③实验数据：... \\ Data \\ 栅格数据 \\ Ex9.gdb：
- DEM_CASE1；
- Soils_Case1；

……

## 四、内容与步骤

本实验实现栅格数据重分类的基本功能：用户右键单击【Geoprocessing】主菜单上【Raster Reclass】菜单项，激活重分类对话框，用户可选择输入图层、输出文件、分类级数、编辑映射表等。

1. RasterReclass 窗体设计

**(1)添加栅格数据重分类对话框类**

新建一个 Windows 窗体，并命名为“RasterReclassFrm”，修改窗体的 Text 属性为“RasterReclass”，并添加 Button、Label、TextBox、ComboBox、DataGridView 控件。控件布局如图 19-1 所示。

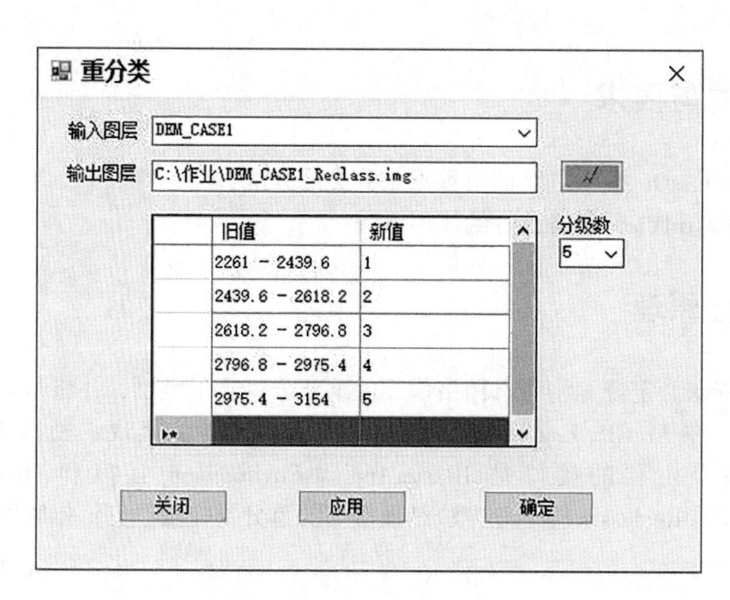


图 19-1　RasterReclassFrm 控件布局

**(2)设置控件属性**

设置相应控件的相关属性，见表 19-1。

表 19-1　**RasterReclassFrm 控件命名表**

| 控件类型 | Name 属性 | 含义 |
| --- | --- | --- |
| ComboBox | cbxInLayers | 输入栅格数据 |
| ComboBox | cbxCount | 分类级数 |
| TextBox | txtOutput | 输出结果文件名 |
| DataGridView | dataGridView1 | 展示映射表 |
| Button | btnBrower | 设置输出文件 |
| Button | btnApp | 应用 |
| Button | btnCancel | 取消 |
| Button | btnOK | 确定 |

**(3) 添加 RasterReclassFrm 的全局变量**

```
private IMapControl3 _mapControl;
private DataTable _DataTable= null;
```

**(4) 添加 RasterReclassFrm 事件响应函数、功能函数、辅助函数**

①添加应用按钮 Click 事件响应函数;

②添加确定按钮 Click 事件响应函数。

代码如下:

```
public partial class RasterReclassFrm : Form
{
    private IMapControl3 _mapControl = null;
    private DataTable _DataTable= null;

    public RasterReclassFrm(IMapControl3 mapControl)
    {
        InitializeComponent();
        _mapControl = mapControl;
        cbxCount.Text = "5";
    }

    //事件响应函数
    private void RasterReclassFrm_Load(object sender, EventArgs e)
    private void btnBrower_Click(object sender, EventArgs e)
    private void cbxInLayers_SelectedIndexChanged(object sender,
EventArgs e)
     private void cbxClassCount _ SelectedIndexChanged ( object
sender, EventArgs e)
    private void btnCancel_Click(object sender, EventArgs e)
    private void btnOK_Click(object sender, EventArgs e)
    private void btnApp_Click(object sender, EventArgs e)

    //辅助函数
    private double[] CreateStretchBreakClass(IRasterLayer raster
Layer, int DesiredClasses)
    private ILayer GetLayerByname(string lyrName)
}
```

### 2. RasterReclassFrm 类的实现

#### (1)Load 事件响应函数的实现

Load 事件响应函数的实现需完成两件事：一是用 DataTable 初始化 DataGridView 表头，二是用 MapControl 中图层名填充 cbxInLayers(输入图层)控件，代码如下：

```
private void RasterReclassFrm_Load(object sender, EventArgs e)
{
    DataColumn OldValue = new DataColumn("旧值", System.Type.
GetType("System.String"));
    DataColumn NewValue = new DataColumn("新值", System.Type.
GetType("System.Int32"));
    _DataTable = new DataTable();
    _DataTable.Columns.Add(OldValue);
    _DataTable.Columns.Add(NewValue);

    dataGridView1.DataSource = _DataTable;

    for (int i = 0;i< _mapControl.LayerCount;i++)
    {
        ILayer pLayer = _mapControl.get_Layer(i);
        if (pLayer is IRasterLayer)
        {
            cbxInLayers.Items.Add(pLayer.Name);
        }
    }
}
```

#### (2)输入图层选择响应函数、分类数选择响应函数

输入图层选择响应函数：利用选定的图层创建分级数组(用到辅助函数 CreateStretchBreakClass())，然后利用分级数组转换为数值范围字符串(×××-×××)填充_DataTable，再更新 DataGradView 数据源。

分类数选择响应函数，只需简单执行输入图层选择响应函数即可。代码如下：

```
private void cbxInLayers_SelectedIndexChanged(object sender,
EventArgs e)
{
    IRasterLayer rasterLayer = GetLayerByname(this.cbxInLayers.
Text) as IRasterLayer;
    if (rasterLayer == null)
        return;
```

```
        //创建分级数组
        int clsCount = Convert.ToInt16(cbxCount.Text);
        double[] dblArr = CreateStretchBreakClass(rasterLayer,
clsCount);

        //填充_DataTable
        _DataTable.Clear();
        for (int i = 1; i < dblArr.Length; i++)
        {
            DataRow pNewRow = _DataTable.NewRow();
            double fvalue = dblArr[i-1];
            double tvalue = dblArr[i];
            pNewRow[0] = fvalue.ToString() + " - " + tvalue.ToString();
            pNewRow[1] = i;
            _DataTable.Rows.Add(pNewRow);
        }

        dataGridView1.DataSource = _DataTable;
        dataGridView1.Refresh();
    }

    private void cbxClassCount_SelectedIndexChanged(object sender,
EventArgs e)
    {
         cbxInLayers_SelectedIndexChanged(sender,  e);
    }
```

**(3)输出文件设置响应函数**

输出路径设置由 SaveFileDialog 实现，代码如下：

```
    private void btnBrower_Click(object sender, EventArgs e)
    {
        SaveFileDialog SaveFileDlg = new SaveFileDialog();
        {
            SaveFileDlg.CheckPathExists = true;
            SaveFileDlg.Filter = "Rasterfile (*.img)|*.img";
            SaveFileDlg.OverwritePrompt = true;
            SaveFileDlg.Title = "Output Layer";
            SaveFileDlg.RestoreDirectory = true;
```

```
            SaveFileDlg.FileName = System.IO.Path.GetFileName
WithoutExtension(cbxInLayers.Text) + "_Reclass.img";
        }
        if (SaveFileDlg.ShowDialog() == DialogResult.OK)
        {
            this.txtOutput.Text = SaveFileDlg.FileName.Trim();
        }
    }
```

**(4)应用响应函数 btnApp_Click()**

应用响应函数执行重分类操作。步骤如下:

①根据 DataGridView 中定义的新旧数值对照关系，构造新旧数值映射图(Remap);

②创建重分类接口对象(IReclassOp)，执行重分类计算;

③结果保存到指定位置;

④添加到 Map。

代码如下:

```
    private void btnApp_Click(object sender, EventArgs e)
    {
        IRasterLayer pLayer = GetLayerByname(this.cbxInLayers.Text)
as IRasterLayer;
        if (pLayer == null)
            return;
        IRaster pRster = pLayer.Raster;

        //1:创建分类映射
        INumberRemap pNumRemap = new NumberRemapClass();
        for(int i = 0 ;i< Convert.ToInt32(cbxCount.Text);i++)
        {
            String str = dataGridView1[0, i].Value.ToString();
            int p = str.LastIndexOf(" - ");
            float fvalue = Convert.ToSingle(str.Substring(0, p));
            float tvalue = Convert.ToSingle(str.Substring(p+3, str.
Length - (p+3) ));
            pNumRemap.MapRange(fvalue, tvalue, i);
        }
        //pNumRemap.MapValueToNoData(-9999)

        //2:执行重分类计算
        IReclassOp reCla = new RasterReclassOpClass() ;
```

```
        IGeoDataset outputRaster = reCla.ReclassByRemap(pRster as
IGeoDataset, pNumRemap as IRemap,true);

        //3:结果保存到指定位置
        string outputFileName = this.txtOutput.Text;
        string outputPath = System.IO.Path.GetDirectoryName(output
FileName);
        string fileName = System.IO.Path.GetFileName(outputFileName);
        IWorkspaceFactory wsf = new RasterWorkspaceFactoryClass();
        IWorkspace ws = wsf.OpenFromFile(outputPath, 0);

        ISaveAs pSaveAs = outputRaster as ISaveAs;
        pSaveAs.SaveAs(fileName, ws, "IMAGINE Image");

        //4:添加到 MapControl
        IRasterLayer pRlayer;
        pRlayer = new RasterLayer();
        pRlayer.CreateFromRaster(outputRaster as Raster);
        pRlayer.Name = fileName + "_Reclass";
        _mapControl.AddLayer(pRlayer,0);
    }

    private void btnCancel_Click(object sender, EventArgs e)
    {
        this.Close();
    }

    private void btnOK_Click(object sender, EventArgs e)
    {
        this.Close();
    }
```

**(5)辅助函数 CreateStretchBreakClass()的实现**

本函数创建分级数组，方法是从 IRasterBand 的 ComputeStatsAndHist()函数获得栅格值的统计数据，然后根据最小值和最大值等间距构造分级数组(注意：为简化学习过程，这里仅用简单的分级方法)，结果取整到小数后第六位，代码如下：

```
    private double[] CreateStretchBreakClass(IRasterLayer raster
Layer, int DesiredClasses)
    {
        //获得栅格数据第一波段
```

```
        IRasterBandCollection pRsBandCol = rasterLayer.Raster as
IRasterBandCollection;
        IRasterBand pRsBand = pRsBandCol.Item(0);

        //获得最大值、最小值,以设置分类级数
        pRsBand.ComputeStatsAndHist();
        IRasterStatistics pRasterStatistic = pRsBand.Statistics;
        double dMaxValue = pRasterStatistic.Maximum;
        double dMinValue = pRasterStatistic.Minimum;

        //构造分级数组
        double BinInterval = Convert.ToDouble((dMaxValue - dMinValue)
/DesiredClasses);
        double[] dblValues = new double[DesiredClasses + 1];
        for (int i = 0; i < DesiredClasses + 1; i++)
        {
            double value = i * BinInterval + dMinValue;
            dblValues[i] = Math.Round(value, 6);
        }

        return dblValues;
    }
```

图层查找辅助函数代码如下:

```
    private ILayer GetLayerByname(string lyrName)
    {
        ILayer pLayer = null;
        for (int i = 0; i < _mapControl.LayerCount; i++)
        {
            ILayer tempLayer = _mapControl.get_Layer(i);
            if (tempLayer.Name == lyrName)
            {
                pLayer = tempLayer;
                break;
            }
        }

        return pLayer;
    }
```

### 3. 功能调用

在主菜单 Geoprocessing 上添加一菜单项(命名为 Raster Reclass)，创建并修改 Click 事件响应函数，代码如下：

```
private void rasterReclassToolStripMenuItem_Click(object sender,
EventArgs e)
{
    RasterReclassFrm frm = new RasterReclassFrm(m_mapControl);
    if (frm.ShowDialog() == DialogResult.OK)
    {
        this.m_mapControl.ActiveView.Refresh();
    }
}
```

## 五、功能测试

①单击 F5 键，编译运行程序。
②加载数据：... \\ Data \\ 栅格数据 \\ Ex9. gdb \\ DEM_CASE1。
③点击菜单【Raster Reclass】，弹出分析窗口，设置输出文件名，选择：
◆ 分析图层=DEM_CASE1；
◆ 分类级数=6；
◆ 确认映射表是否符合要求(必要时做适当编辑)。
④单击【应用】即生成重新分类后的数据。
最终效果如图 19-2 所示。

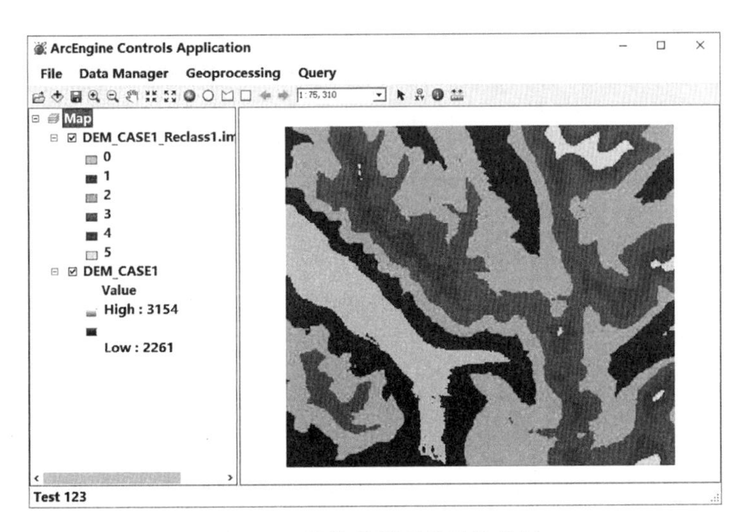


图 19-2　栅格数据重分类效果图

## 六、思考与练习

试使用“自然断点法”进行重分类。

# 实验二十　栅格计算

## 一、目的与要求

①掌握栅格分析环境设置方法；
②掌握使用 lMathOp、ILogicalOp 进行栅格计算的一般步骤；
③深入理解 IRasterDataset、IRaster 接口各自的含义。

## 二、目的与要求

栅格计算是栅格数据空间分析中最为常用的方法，利用栅格计算，可将复杂的矢量数据叠加分析转变为简单的算术运算。ArcGIS Engine 支持数学计算、三角函数、逻辑运算和按位运算等栅格计算类型。RasterMathOpsClass 实现了所有的栅格计算接口，对应于以上的几种计算类型，该类实现了如下接口：

①lMathOp，包含数学计算的所有方法：加(Plus)、减(Minus)、乘(Times)、除(Divide)、绝对值(Abs)、指数(Exp)和对数(Ln)等；

②ILogicalOp，包含逻辑运算的所有方法：布尔与(Boolean And)、布尔非(Boolean Not)、布尔或(Boolean Or)、大于(Greater Than)、大于等于(GreaterThan Equal)、小于(Less Than)、小于等于(LessThan Equal) 和等于(Equal To)等；

③ITrigOp，包含三角函数运算的所有方法；

④IBitwiseOp，包含按位运算的所有方法；

⑤IRaster AnalysisEnvironment，设置空间分析环境。

## 三、实验环境

①开发环境：Visual Studio 2015 + ArcGIS Engine 10.5。
②开发语言：C#。
③实验数据：… \\ Data \\ 栅格数据 \\ Ex9. gdb：
◆ DEM_CASE1；
◆ Soils_Case1。

## 四、内容与步骤

本实验实现栅格计算的基本功能：用户右击【Geoprocessing】主菜单上【Raster Caculator】菜单项，激活栅格数据计算对话框，用户可选择输入图层1，输入图层2、输出文件、计算方法等。

### 1. Raster Caculator 窗体设计

#### (1)添加栅格计算对话框类

新建一个 Windows 窗体，并命名为“RasterCaculatorFrm”，修改窗体的 Text 属性为“Raster Caculator”，并添加 Button、Label、TextBox，Combobox 控件。控件布局如图 20-1 所示。

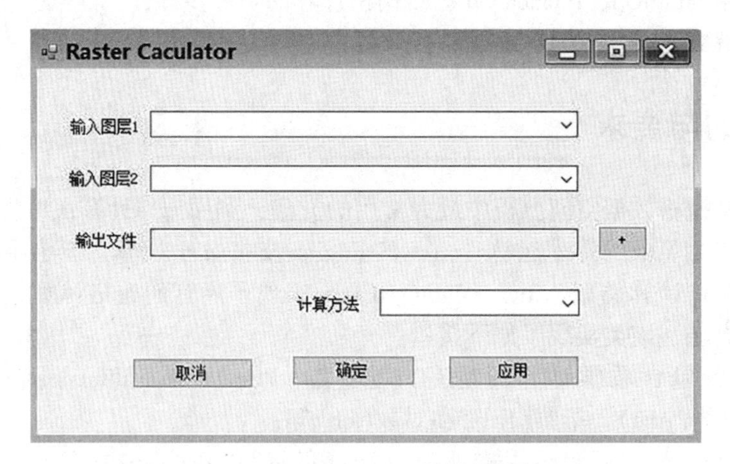


图 20-1　Raster Caculator 窗体控件布局

#### (2)设置控件属性

设置相应控件的相关属性，见表 20-1。

表 20-1　**Raster Caculator 窗体控件命名表**

| 控件类型 | Name 属性 | 含义 | 备注 |
|---|---|---|---|
| ComboBox | cbxRasterLyr1 | 输入栅格图层 1 | |
| ComboBox | cbxRasterLyr2 | 输入栅格图层 2 | 也可输入一个常数 |
| ComboBox | cbxCaculateMethod | 计算方法 | |
| TextBox | txtOutputFile | 输出结果文件名 | |
| Button | btnBrower | 设置输出文件 | |
| Button | btnApp | 应用 | |
| Button | btnCancel | 取消 | |
| Button | btnOK | 确定 | |

**(3)添加 RasterCaculatorFrm 的全局变量**

```
private IMapControl3 _mapControl;
```

**(4)添加 RasterReclassFrm 事件响应函数、功能函数、辅助函数**

①添加应用按钮 Click 事件响应函数。

②添加确定按钮 Click 事件响应函数。

代码如下：

```
public partial class RasterCaculatorFrm : Form
{
    private IMapControl3 _mapControl = null;
    public RasterCaculatorFrm(IMapControl3 mapControl)
    {
        InitializeComponent();
        _mapControl = mapControl;
    }

    //事件响应函数
    private void RasterCaculatorFrm_Load(object sender, EventArgs e)
    private void btnBrower_Click(object sender, EventArgs e)
    private void btnOK_Click(object sender, EventArgs e)
    private void btnCancel_Click(object sender, EventArgs e)
    private void btnApp_Click(object sender, EventArgs e)

    //功能函数
    private IGeoDataset Caculate(IRaster pIRaster1, IRaster
pIRaster2, string strCalculateMethod)

    //辅助函数
    private void SetAnalysisEnvironment(IRasterAnalysis Environment
rasAnaEnv, IRasterProps rProps)
    public static IRaster MakeConstantRaster(string path, IRaster
Props rProps, double value)
    private IRasterLayer GetRasterLayer(string layerName)
    private IEnumLayer GetRasterLayers()
}
```

## 2. RasterCaculatorFrm 类的实现

**(1)Load 事件响应函数的实现**

用 MapControl 中图层名填充 cbxRasterLyr1、cbxRasterLyr2(输入图层)控件，用加、

减、乘、除、大于、小于字符填充 cbxCalculateMethod，设置默认输出文件名。

代码如下：

```
private void RasterCaculatorFrm_Load(object sender, EventArgs e)
{
    IEnumLayer layers = GetRasterLayers();
    layers.Reset();
    ILayer layer = null;
    while ((layer = layers.Next()) ! = null)
    {
        cbxRasterLyr1.Items.Add(layer.Name);
        cbxRasterLyr2.Items.Add(layer.Name);
    }

    this.cbxCalculateMethod.Items.Add("加(Plus)");
    this.cbxCalculateMethod.Items.Add("减(Minus)");
    this.cbxCalculateMethod.Items.Add("乘(Times)");
    this.cbxCalculateMethod.Items.Add("除(Divide)");
    this.cbxCalculateMethod.Items.Add("大于(Greater Than)");
    this.cbxCalculateMethod.Items.Add("小于(Less Than)");
    this.cbxCalculateMethod.SelectedIndex = 0;

    //select the first layer
    if (cbxRasterLyr1.Items.Count > 0)
        cbxRasterLyr1.SelectedIndex = 0;
    if (cbxRasterLyr2.Items.Count > 0)
        cbxRasterLyr2.SelectedIndex = 0;

    string tempDir = System.IO.Path.GetTempPath();
    string strFileName =
                        System.IO.Path.GetFileNameWithoutExtension
(cbxRasterLyr1.Text);
    strFileName += "_" + cbxCalculateMethod.Text + ".img";
    txtOutputFile.Text = System.IO.Path.Combine(tempDir,
strFileName);
}
```

**(2)输出文件设置响应函数**

输出文件设置由 SaveFileDialog 实现，代码如下：

```
private void btnBrower_Click(object sender, EventArgs e)
```

```
{
    //set the output layer
    SaveFileDialog saveDlg = new SaveFileDialog();
    saveDlg.CheckPathExists = true;
    saveDlg.Filter = "IMAGINE Image ( * .img) |* .img";
    saveDlg.OverwritePrompt = true;
    saveDlg.Title = "Output Layer";
    saveDlg.RestoreDirectory = true;
    saveDlg.FileName =
                        System.IO.Path.GetFileNameWithoutExtension
(cbxRasterLyr1.Text);
    saveDlg.FileName += "_"+cbxCalculateMethod.Text + ".img";

    DialogResult dr = saveDlg.ShowDialog();
    if (dr == DialogResult.OK)
        txtOutputFile.Text = saveDlg.FileName;
}
```

**(3)应用响应函数 btnApp_Click( )**

应用响应函数执行栅格计算操作，具体步骤如下：

①获取输入栅格图层 1，输入栅格图层 2。如果输入栅格图层 2 为 null，则将图名转换为常量，并以此常量构造一个常数栅格矩阵(用到辅助函数 MakeConstantRaster)。

②调用核心函数 Caculate 执行栅格计算。

③结果保存到指定位置。

④添加到 Map。

代码如下：

```
private void btnApp_Click(object sender, EventArgs e)
{
    //输入栅格数据 1
    IRasterLayer pRasterLayer1 = GetRasterLayer(cbxRasterLyr1.
Text);
    IRaster pIRaster1 = null;
    if (pRasterLayer1 ! = null)
        pIRaster1 = pRasterLayer1.Raster;
    else
        return;

    //输入栅格数据 2
    IRasterLayer pRasterLayer2 = GetRasterLayer(cbxRasterLyr2.
```

```
Text);;
        IRaster pIRaster2 = null;
        if( pRasterLayer2 ! = null )
            pIRaster2 = pRasterLayer2.Raster;
        else
        {
            string path = System.IO.Path.GetTempPath();
            double value = double.Parse(cbxRasterLyr2.Text);
            pIRaster2 = MakeConstantRaster(path, pIRaster1 as IRaster
Props, value);
        }

        //栅格计算
        string strCalculateMethod = this.cbxCalculateMethod.Text;
        IGeoDataset p3 = Caculate(pIRaster1, pIRaster2, strCalculate
Method);

        //结果保存到指定位置
        string outputFileName = this.txtOutputFile.Text;
        string outputPath = System.IO.Path.GetDirectoryName
(outputFileName);
        string fileName = System.IO.Path.GetFileName
(outputFileName);
        IWorkspaceFactory wsf = new RasterWorkspaceFactoryClass();
        IWorkspace ws = wsf.OpenFromFile(outputPath, 0);

        ISaveAs pSaveAs = p3 as ISaveAs;
        pSaveAs.SaveAs(fileName, ws, "IMAGINE Image");

        //添加到 MapControl
        IRasterLayer pRlayer;
        pRlayer = new RasterLayer();
        pRlayer.CreateFromRaster(p3 as Raster);
        pRlayer.Name = fileName;
        _mapControl.AddLayer(pRlayer, 0);
    }

    private void btnCancel_Click(object sender, EventArgs e)
```

```
    {
        this.Close();
    }

    private void btnOK_Click(object sender, EventArgs e)
    {
        this.Close();
    }
```

**(4)核心函数 Caculate( )的实现**

核心函数 Caculate( )完成栅格叠加计算，先初始化 RasterMathOpsClass，设置栅格分析环境，然后根据计算方法执行相应的栅格计算。

代码如下：

```
    private IGeoDataset Caculate(IRaster pIRaster1, IRaster pIRaster2,
string strCalculateMethod)
    {
        //初始化栅格计算类
        IMathOp pMathOp = new RasterMathOpsClass();
        //设置栅格分析环境
        SetAnalysisEnvironment(pMathOp as IRasterAnalysisEnvironment,
pIRaster1 as IRasterProps);

        IGeoDataset p1 = pIRaster1 as IGeoDataset;
        IGeoDataset p2 = pIRaster2 as IGeoDataset;
        IGeoDataset p3 = null;
        //根据计算方法执行计算
        switch (strCalculateMethod)
        {
            case "加(Plus)":
                p3 = pMathOp.Plus(p1, p2);
                break;
            case "减(Minus)":
                p3 = pMathOp.Minus(p1, p2);
                break;
            case "乘(Times)":
                p3 = pMathOp.Times(p1, p2);
                break;
            case "除(Divide)":
                p3 = pMathOp.Divide(p1, p2);
```

```
                    break;
                case "大于(Greater Than)":
                    {
                        ILogicalOp logical = pMathOp as ILogicalOp;
                        p3 = logical.GreaterThan(p1, p2);
                    }
                    break;
                case "小于(Less Than)":
                    {
                        ILogicalOp logical = pMathOp as ILogicalOp;
                        p3 = logical.LessThan(p1, p2);
                    }
                    break;
                default:
                    break;
            }

            return p3;
        }
```

**(5)辅助函数 SetAnalysisEnvironment( )的实现**

辅助函数 SetAnalysisEnvironment( )主要设置范围、栅格元大小，代码如下：

```
    private void SetAnalysisEnvironment(IRasterAnalysisEnvironment
rasAnaEnv,IRasterProps rProps)
    {
        object extent = rProps.Extent;
        IPnt p = rProps.MeanCellSize();
        object cellsize = (p.X + p.Y) /2;
        object missing = Type.Missing;

        //设置生成图层的范围
        rasAnaEnv.SetExtent(esriRasterEnvSettingEnum.esriRaster
EnvValue, ref extent, ref missing);
        //设置生成图层的栅格大小
        rasAnaEnv.SetCellSize(esriRasterEnvSettingEnum.esriRaster
EnvMinOf, ref cellsize);
    }
```

MakeConstantRaster( )函数可创建常数栅格矩阵，主要用 RasterMakerOpClass 的 IRasterMakerOp 接口实现，代码如下：

```
    public IRaster MakeConstantRaster ( string path, IRasterProps
rPops, double value)
    {
        IWorkspaceFactory workspaceFactory = new RasterWorkspace
FactoryClass();
        IWorkspace workspace = workspaceFactory.OpenFromFile(path,
0);

        IRasterMakerOp rasterMakerOp = new RasterMakerOpClass();
        IRasterAnalysisEnvironment rasAnaEnv = (IRasterAnalysis
Environment)rasterMakerOp;
        {
            SetAnalysisEnvironment(rasAnaEnv, rProps);
            rasAnaEnv.OutWorkspace = workspace;
        }

        IRaster raster = rasterMakerOp.MakeConstant(value, true) as
IRaster;
        return raster;
    }
```

GetRasterLayer()，GetRasterLayers()两个函数，依据图层面取得栅格数据图层接口，代码如下：

```
    private IRasterLayer GetRasterLayer(string layerName)
    {
        //get the layers from the maps
        IEnumLayer layers = GetRasterLayers();
        layers.Reset();

        ILayer layer = null;
        while ((layer = layers.Next()) ! = null)
        {
            if (layer.Name == layerName)
                return layer as IRasterLayer;
        }

        return null;
    }
```

```
private IEnumLayer GetRasterLayers()
{
    UID uid = new UIDClass();
    //uid.Value = "{40A9E885-5533-11d0-98BE-00805F7CED21}";//
FeatureLayer
    uid.Value = "{D02371C7-35F7-11D2-B1F2-00C04F8EDEFF}"; //
RasterLayer
    IEnumLayer layers = _mapControl.Map.get_Layers(uid, true);
    return layers;
}
```

### 3. 功能调用

在主菜单【Geoprocessing】上添加一菜单项(命名为“Raster Caculator”)，创建并修改Click事件响应函数，代码如下：

```
private void rasterCaculateToolStripMenuItem_Click(object sender,
EventArgs e)
{
    RasterCaculatorFrm frm = new RasterCaculatorFrm(m_
mapControl);
    if (frm.ShowDialog() == DialogResult.OK)
    {
        this.m_mapControl.ActiveView.Refresh();
    }
}
```

## 五、功能测试

①单击F5键，编译运行程序。

②点击菜单【Raster Caculator】，弹出分析窗口，设置输出文件名，选择：

- 分析图层1= DEM_CASE1;
- 输入分析图层2= 2500(为常数);
- 分析方法= Greater Than;

③单击【应用】按钮即生成新的栅格数据。

最终效果如图20-2所示。

## 六、思考与练习

设置栅格分析环境有何意义?

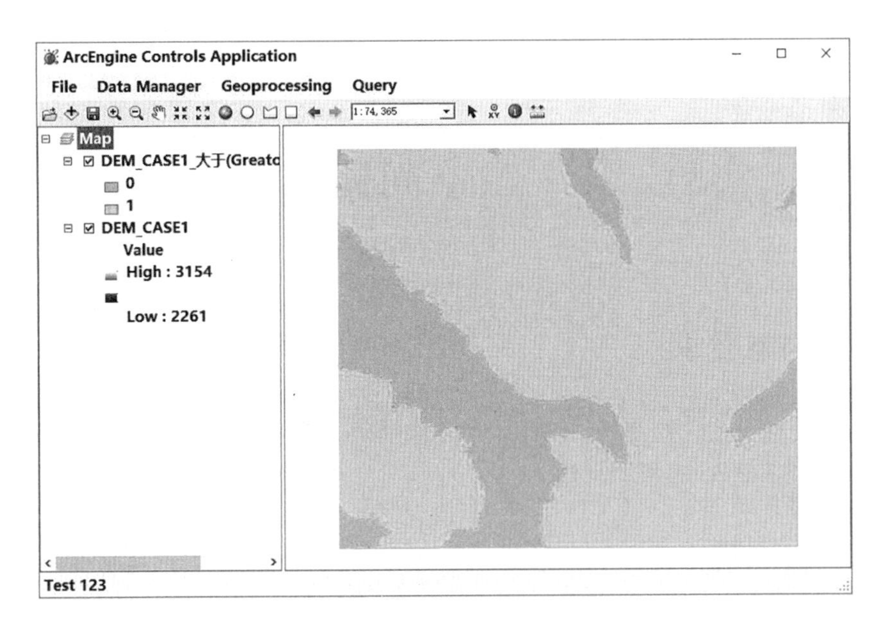


图 20-2　栅格计算效果图

# 实验二十一　空间插值(IDW)

## 一、目的与要求

通过对反距离权重法的使用，熟悉了解空间插值方法的使用。有关其他插值方法，读者可参阅 ArcGIS Engine 的帮助文档。

## 二、实验原理

空间插值是根据有限的样本点数据来生成连续的表面(像素分布)，根据插值点计算公式的不同可分为多种插值方法，其中最常用的是 IDW 法(反距离加权法)，它以插值点与样本点之间的距离倒数为权重进行加权平均，离插值点越近的样本点赋予的权重越大。

在 ArcGIS Engine 中，RasterlnterpolationOpClass 类专用于空间插值分析。该类实现了两个主要接口，分别是 IRasterAnalysisEnvironment 接口和 IInterpolationOp 接口。

IRasterAnalysisEnvironment 接口用来设置空间分析的环境，IInterpolationOp 接口可实现所有空间插值的方法，分别为：

①反距离权重法(IDW)；

②克里金法(Kriging)；

③样条函数法(Spline)；

④趋势面法(Trend)；

⑤自然邻域法(NaturalNeighbor)；

⑥通过文件实现地形转栅格(TopoToRasterByFile)；

⑦变异函数法(Variogram)。

## 三、实验环境

①开发环境：Visual Studio 2015 + ArcGIS Engine 10.5。

②开发语言：C#。

③实验数据：... \\ Data \\ 栅格数据 \\ GDP.gdb：

- ◆ jsGDP；
- ◆ jsGDP_training。

## 四、内容与步骤

本实验实现 IDW 空间插值的功能：用户右击【Geoprocessing】主菜单上【IDW Interpolation】菜单项，激活 IDW 空间插值对话框，用户可选择输入图层、输出文件，以及控制参数表等。

1. IDWInterpolation 窗体设计

**(1)添加 IDW 空间插值对话框类**

新建一个 Windows 窗体，并命名为“IDWInterpolationFrm”，修改窗体的 Text 属性为“IDW 插值”，并添加 Button、Label、ComboBox 等控件。控件布局如图 21-1 所示。

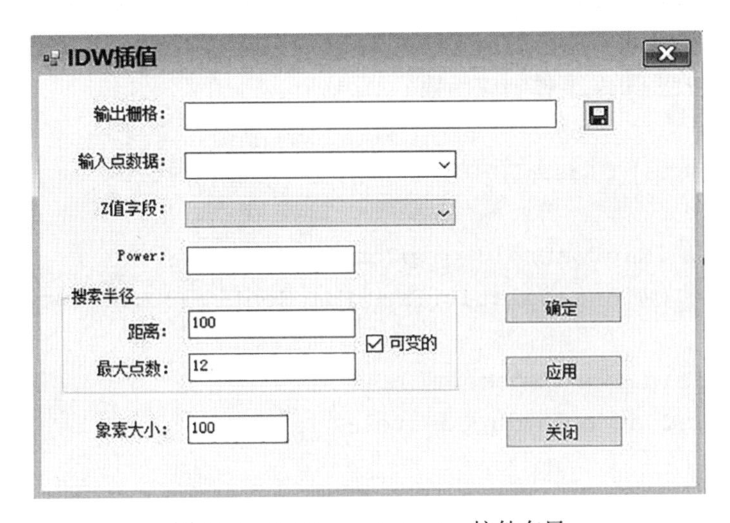


图 21-1　IDWInterpolationFrm 控件布局

**(2)设置控件属性**

设置相应控件的相关属性，见表 21-1。

表 21-1　**IDWInterpolationFrm 控件命名表**

| 控件 | Name | 含义 | 备注 |
| --- | --- | --- | --- |
| TextBox | txtOutput | 输出文件名 | Img 格式 |
| ComboBox | cbxInPointLyr | 输入点图层 | |
| ComboBox | cbxZValueField | Z 值字段 | |
| TextBox | txtPowerValue | 权重指数 | |
| TextBox | txtMaxDistance | 最大搜索距离 | |
| TextBox | txtMinPoints | 最少搜索点数 | |

续表

| 控件 | Name | 含义 | 备注 |
|---|---|---|---|
| TextBox | txtCellSize | 像素大小 | |
| Button | btnExplore | 文件浏览 | |
| Button | btnApp | 应用 | |
| Button | btnCancel | 关闭 | |
| Button | btnOK | 确定 | |

**(3)添加 IDWInterpolationFrm 的全局变量**

```
private IMapControl3 _mapControl;
```

**(4)添加 IDWInterpolationFrm 事件响应函数、功能函数、辅助函数**

①添加应用按钮 Click 事件响应函数。

②添加确定按钮 Click 事件响应函数。

代码如下：

```
public partial class IDWInterpolationFrm : Form
{
    private IMapControl3 _mapControl = null;
    public IDWInterpolationFrm(IMapControl3 mapControl)
    {
        InitializeComponent();
        _mapControl = mapControl;
    }

    //事件响应函数
        private void IDWInterpolationFrm_Load(object sender,
EventArgs e)
        private void cbxInPointLyr_SelectedIndexChanged(object
sender, EventArgs e)
        private void btnExplore_Click(object sender, EventArgs e)
        private void btnApp_Click(object sender, EventArgs e)
        private void btnOK_Click(object sender, EventArgs e)
        private void btnCancel_Click(object sender, EventArgs e)

    //辅助函数
        private void SetRasterAnalysisEnvironment(IInterpolationOp
pInterpolationOp, IFeatureLayer maskFlyr, IEnvelope pExtent, double
cellsize)
```

```
        private IRasterRadius GetSearchRadius()
        private void SaveRasterToFile(IRaster pRaster, String sPath,
String sOutName)
        private ILayer GetLayerByname(string lyrName)
    }
```

### 2. IDWInterpolationFrm 类的实现

**(1) Load 事件响应函数的实现**

Load 事件响应函数的主要作用是：用 MapControl 中点类图层名填充 cbxInPointLyr(图层名)控件，同时对几个分析参数赋初值。代码如下：

```
private void IDWInterpolationFrm_Load(object sender, EventArgs e)
{
    for (int i = 0; i < _mapControl.Map.LayerCount; i++)
    {
        ILayer aLayer = _mapControl.Map.get_Layer(i);
        if (aLayer is IFeatureLayer)
        {
            IFeatureLayer flyr = (IFeatureLayer)aLayer;
            if (flyr.FeatureClass.ShapeType == esriGeometryType.
esriGeometryPoint)
                this.cbxInPointLyr.Items.Add(aLayer.Name);
        }
    }

    txtOutput.Text = "";
    txtCellSize.Text = "10.0";

    cbxZValueField.Text = "无";
    txtPowerValue.Text = "2";
    txtMinPoints.Text = "12";
}
```

**(2) btnExplore 按钮响应函数的实现**

代码如下：

```
private void btnExplore_Click(object sender, EventArgs e)
{
    SaveFileDialog SaveFileDlg = new SaveFileDialog();
    {
        SaveFileDlg.CheckPathExists = true;
```

```
            SaveFileDlg.Filter = "Rasterfile (*.img)|*.img";
            SaveFileDlg.OverwritePrompt = true;
            SaveFileDlg.Title = "Output Layer";
            SaveFileDlg.RestoreDirectory = true;
            SaveFileDlg.FileName = cbxInPointLyr.Text + "_IDW.img";
        }

        if (SaveFileDlg.ShowDialog() == DialogResult.OK)
        {
            this.txtOutput.Text = SaveFileDlg.FileName.Trim();
        }
    }
```

**(3) cbxInPointLyr_SelectedIndexChanged 图层选择响应函数**

本函数在 cbxInPointLyr 控件选择某图层后，完成对可选 z 值字段的重新填充。代码如下：

```
    private void cbxInPointLyr_SelectedIndexChanged(object sender,
EventArgs e)
    {
        IFeatureLayer pFeatLyr = GetLayerByname(cbxInPointLyr.Text)
as IFeatureLayer;
        if (pFeatLyr == null)
            return;

        cbxZValueField.Items.Clear();
        cbxZValueField.Items.Add("无");
        IFields pFields = pFeatLyr.FeatureClass.Fields;
        for (int j = 0; j < pFields.FieldCount; j++)
        {
            IField pFiled = pFields.get_Field(j);
            switch (pFiled.Type)
            {
                case esriFieldType.esriFieldTypeDouble:
                case esriFieldType.esriFieldTypeSingle:
                case esriFieldType.esriFieldTypeSmallInteger:
                case esriFieldType.esriFieldTypeInteger:
                    cbxZValueField.Items.Add(pFiled.Name);
                    break;
                default:
```

```
                break;
            }
        }
    }
```

**(4)btnApp 按钮响应函数的实现**

btnApp 按钮响应函数完成空间内插操作，步骤如下：

第一步，利用输入要素类构造一个要素描述器 FeatureClassDescriptor；由它定义 FeatureClass 作为分析操作的属性；

第二步，获取分析参数(Power、ZValue、CellSize、RasterRadius)；

第三步，初始化内插计算接口，并设置分析环境；

第四步，调用 IDW 方法，执行内插计算；

第五步，结果保存到文件；

第六步，显示。

代码如下：

```
private void btnApp_Click(object sender, EventArgs e)
{
    IFeatureLayer pntLyr = GetLayerByname(cbxInPointLyr.Text) as
IFeatureLayer;
    IFeatureClass pInPointFClass = pntLyr.FeatureClass;

    //构造一个要素描述器 FeatureClassDescriptor
    IFeatureClassDescriptor pFeatClsDes = new FeatureClass
DescriptorClass() ;
    if (cbxZValueField.Text ! = "无")
        pFeatClsDes.Create(pInPointFClass, null, cbxZValueField.
Text);
    else
        pFeatClsDes.Create(pInPointFClass, null, "");

    //获取分析参数
    double dCellSize = Convert.ToDouble(txtCellSize.Text);
    double dPower = Convert.ToDouble(txtPowerValue.Text);
    IRasterRadius pRsRadius = GetSearchRadius();

    //创建内插计算接口,并设置分析环境
    IInterpolationOp pInterpolationOp = new RasterInterpolation
Op() as IInterpolationOp;
    SetRasterAnalysisEnvironment(pInterpolationOp, null, pntLyr.
```

```
AreaOfInterest, dCellSize);

        //执行内插计算
        object objLineBarrier = Type.Missing;
        IGeoDataset rasDataset;
        rasDataset = pInterpolationOp.IDW((IGeoDataset)pFeatClsDes,
dPower, pRsRadius, ref objLineBarrier);
        IRaster outputRaster = rasDataset as IRaster;

        //保存
        string outputFileName = txtOutput.Text;
        string outputPath = System.IO.Path.GetDirectoryName
(outputFileName);
        string fileName = System.IO.Path.GetFileName(outputFile
Name);
        SaveRasterToFile(outputRaster, outputPath, fileName);

        //显示
        IRasterLayer pRasterLayer = new RasterLayer() as IRaster
Layer;
        pRasterLayer.Name = fileName;
        pRasterLayer.CreateFromRaster(outputRaster);
        _mapControl.AddLayer(pRasterLayer, 0);
    }

    private void btnCancel_Click(object sender, EventArgs e)
    {
        this.Close();
    }

    private void btnOK_Click(object sender, EventArgs e)
    {

    }
```

**(5)辅助函数**

SetRasterAnalysisEnvironment()函数可实现栅格环境设置，包括设置像素大小、分析范围、插值边界(有效范围)，具体代码如下：

```
    private  void  SetRasterAnalysisEnvironment ( IInterpolationOp
```

```
pInterpolationOp,
        IFeatureLayer maskFlyr, IEnvelope pExtent, double cellsize)
    {
         IRasterAnalysisEnvironment pRsEnv = pInterpolationOp as
IRasterAnalysisEnvironment;
        {
            //设置像素大小
            object objCellSize = cellsize;
            pRsEnv.SetCellSize(esriRasterEnvSettingEnum.esriRaster
EnvValue, ref objCellSize);

            //设置插值边界
            if (maskFlyr ! = null)
            {
                IGeoDataset pmaskGeoDB = maskFlyr.FeatureClass as IGeo
Dataset;
                pRsEnv.Mask = pmaskGeoDB;
            }

            //设置范围
            object objExtent = pExtent;
            object Missing = Type.Missing;
            pRsEnv.SetExtent(esriRasterEnvSettingEnum.esriRaster
EnvValue, ref objExtent, ref Missing);
        }
    }
```

GetSearchRadius()函数获取搜索半径，分为“可变”(搜索半径可变，固定至少点数)、“固定”(搜索半径固定，点数不限)两种情况，由 chkVariate 控件控制。代码如下：

```
    //获取搜索半径
    private IRasterRadius GetSearchRadius()
    {
        IRasterRadius pRsRadius = new RasterRadius() as IRaster
Radius;
        {
            int iPointNums = Convert.ToInt32(txtMinPoints.Text);
            double dDistance = Convert.ToDouble(txtMaxDistance.
Text);
```

```
            if (this.chkVariate.Checked)
            {
                object objMaxDis = Type.Missing;
                pRsRadius.SetVariable(iPointNums, ref objMaxDis);
            }
            else
            {
                object objMinPointNums = Type.Missing;
                pRsRadius.SetFixed(dDistance, ref objMinPointNums);
            }
        }

        return pRsRadius;
    }
```

SaveRasterToFile( )函数将结果保存到文件，代码如下：

```
    private void  SaveRasterToFile ( IRaster pRaster, String sPath,
String sOutName)
    {
        if (File.Exists(sPath + "\\" + sOutName) = = true)
            File.Delete(sPath + "\\" + sOutName);

        IWorkspaceFactory pWSF = new RasterWorkspaceFactory ( ) as
IWorkspaceFactory;
        IWorkspace pRWS   = pWSF.OpenFromFile(sPath, 0);

        IRasterBandCollection pRasBandCol = pRaster as IRaster
BandCollection;
        IDataset pDS = pRasBandCol.SaveAs ( sOutName, pRWS, " IMAGINE
Image");

        ITemporaryDataset pRsGeo  = (ITemporaryDataset)pDS;
        if (pRsGeo.IsTemporary())
            pRsGeo.MakePermanent();
    }
```

GetLayerByname( )函数根据图层名获取图层，代码如下：

```
    private ILayer GetLayerByname(string lyrName)
```

```
{
    ILayer pLayer = null;
    for (int i = 0; i < _mapControl.LayerCount; i++)
    {
        ILayer tempLayer = _mapControl.get_Layer(i);
        if (tempLayer.Name == lyrName)
        {
            pLayer = tempLayer;
            break;
        }
    }

    return pLayer;
}
```

### 3. 功能调用

在主菜单【Geoprocessing】上添加一菜单项(命名为“Interpolate IDW”)，创建并修改Click事件响应函数，代码如下：

```
private void interpolateIDWToolStripMenuItem_Click(object sender, EventArgs e)
{
    IDWInterpolationFrm frm = new IDWInterpolationFrm(m_mapControl);
    if (frm.ShowDialog() == DialogResult.OK)
    {
        m_mapControl.ActiveView.Refresh();
        axTOCControl1.Update();
    }
}
```

## 五、功能测试

①单击 F5 键，编译运行程序。

②加载数据:… \\ Data \\ 栅格数据 \\ GDP. gdb \\ jsGDP。

③点击菜单【Interpolate IDW】，启动要素类创建对话框，设置输出的文件名，选择：

- 输入图层 = jsGDP;
- Z 值字段 = POPULATION;
- 搜索方法 = 可变;

◆ 最少点数= 12;

◆ 像素大小= 500;

④单击【应用】按钮，即可内插栅格数据到已添加地图。

最终效果如图 21-2 所示。

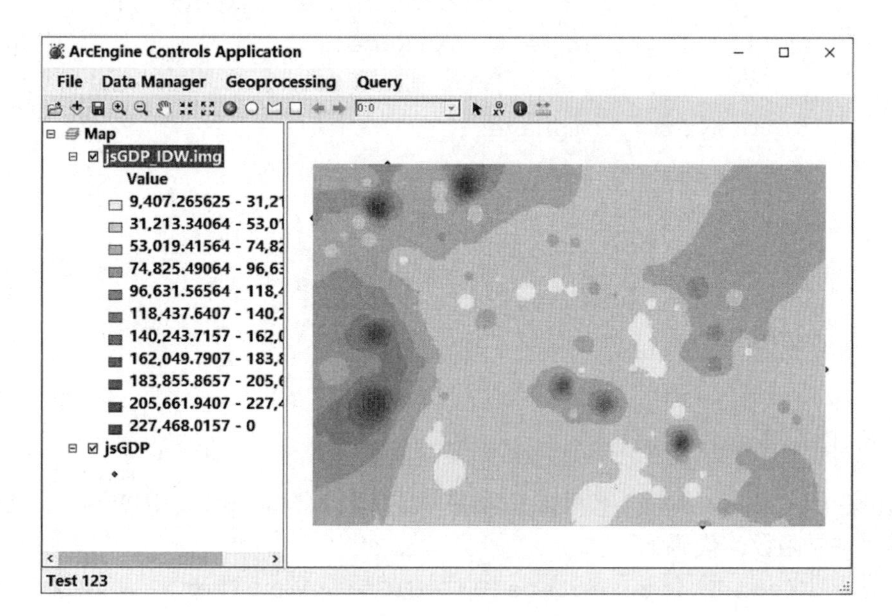


图 21-2 空间插值效果图

## 六、思考与练习

①尝试实现样条空间插值分析功能。

②尝试实现克里金空间插值分析功能。

# 附录 1　LicenseInitializer 源代码

```
internal sealed class LicenseInitializer
{
    private IAoInitialize m_AoInit = new AoInitializeClass();

    #region Private members
    private const string MessageNoLicensesRequested = "Product: No
licenses were requested";
    private const string MessageProductAvailable = "Product: {0}:
Available";
    private const string MessageProductNotLicensed = "Product: {0}:
Not Licensed";
    private const string MessageExtensionAvailable = " Extension:
{0}: Available";
    private const string MessageExtensionNotLicensed = " Extension:
{0}: Not Licensed";
    private const string MessageExtensionFailed = " Extension: {0}:
Failed";
    private const string MessageExtensionUnavailable = " Extension:
{0}: Unavailable";

    private bool m_hasShutDown = false;
    private bool m_hasInitializeProduct = false;

    private List<int> m_requestedProducts;
    private List<esriLicenseExtensionCode> m_requestedExtensions;
    private Dictionary<esriLicenseProductCode, esriLicenseStatus> m_
productStatus = new
    Dictionary<esriLicenseProductCode, esriLicenseStatus>();
    private Dictionary<esriLicenseExtensionCode, esriLicenseStatus>
m_extensionStatus = new
    Dictionary<esriLicenseExtensionCode, esriLicenseStatus>();
```

```
    private bool m_productCheckOrdering = true; //default from low to
high
    #endregion

     public bool InitializeApplication ( esriLicenseProductCode [ ]
productCodes,
    esriLicenseExtensionCode[] extensionLics)
    {
            //Cache product codes by enum int so can be sorted without
custom sorter
            m_requestedProducts = new List<int>();
            foreach (esriLicenseProductCode code in productCodes)
            {
            int requestCodeNum = Convert.ToInt32(code);
            if (! m_requestedProducts.Contains(requestCodeNum))
            {
                m_requestedProducts.Add(requestCodeNum);
            }
            }

            AddExtensions(extensionLics);
            return Initialize();
    }

    ///<summary>
    ///A summary of the status of product and extensions initialization.
    ///</summary>
    public string LicenseMessage()
    {
            string prodStatus = string.Empty;
            if (m_productStatus == null ||m_productStatus.Count== 0)
            {
                prodStatus = MessageNoLicensesRequested+Environment.
NewLine;
            }
            else if (m_productStatus.ContainsValue(esriLicenseStatus.
esriLicenseAlreadyInitialized)
                    ||m_productStatus.ContainsValue(esriLicenseStatus.
```

```
esriLicenseCheckedOut))
            {
                prodStatus = ReportInformation(m_AoInit as ILicense
Information,
                m_AoInit.InitializedProduct(),
                esriLicenseStatus.esriLicenseCheckedOut) + Environment.
NewLine;
            }
            else
            {
                //Failed...
                foreach (KeyValuePair<esriLicenseProductCode,
esriLicenseStatus> item in m_productStatus)
                {
                    prodStatus += ReportInformation(m_AoInit as ILicense
Information,
                    item.Key, item.Value) + Environment.NewLine;
                }
            }

            string extStatus = string.Empty;
            foreach (KeyValuePair<esriLicenseExtensionCode,
esriLicenseStatus> item in m_extensionStatus)
            {
                string info = ReportInformation(m_AoInit as ILicense
Information, item.Key, item.Value);
                if (! string.IsNullOrEmpty(info))
                    extStatus += info + Environment.NewLine;
            }

            string status = prodStatus + extStatus;
            return status.Trim();
        }
        ///<summary>
        ///Shuts down AoInitialize object and check back in extensions to
ensure
        /// any ESRI libraries that have been used are unloaded in the
correct order.
```

```
    ///</summary>
    ///<remarks>Once Shutdown has been called, you cannot re-
initialize the product license
    ///and should not make any ArcObjects call.
    ///</remarks>
    public void ShutdownApplication()
    {
        if (m_hasShutDown)
            return;

        //Check back in extensions
        foreach (KeyValuePair<esriLicenseExtensionCode, esri
LicenseStatus> item in m_extensionStatus)
        {
            if (item.Value == esriLicenseStatus.esriLicense
CheckedOut)
                m_AoInit.CheckInExtension(item.Key);
        }

        m_requestedProducts.Clear();
        m_requestedExtensions.Clear();
        m_extensionStatus.Clear();
        m_productStatus.Clear();
        m_AoInit.Shutdown();
        m_hasShutDown = true;
        //m_hasInitializeProduct = false;
    }

    ///<summary>
    ///Indicates if the extension is currently checked out.
    ///</summary>
    public bool IsExtensionCheckedOut(esriLicenseExtensionCode
code)
    {
        return m_AoInit.IsExtensionCheckedOut(code);
    }

    ///<summary>
```

```
    ///Set the extension(s) to be checked out for your ArcObjects code.
    ///</summary>
    public bool AddExtensions (params esriLicenseExtensionCode[]
requestCodes)
    {
            if (m_requestedExtensions == null)
                m_requestedExtensions = new List<esriLicense
ExtensionCode>();
            foreach (esriLicenseExtensionCode code in requestCodes)
            {
                if (! m_requestedExtensions.Contains(code))
                    m_requestedExtensions.Add(code);
            }

            if (m_hasInitializeProduct)
                    return CheckOutLicenses(this.InitializedProduct);

            return false;
    }

    ///<summary>
    ///Check in extension(s) when it is no longer needed.
    ///</summary>
    public void RemoveExtensions(params esriLicenseExtensionCode[]
requestCodes)
    {
            if (m_extensionStatus == null ||m_extensionStatus.
Count == 0)
            return;

            foreach (esriLicenseExtensionCode code in requestCodes)
            {
                if (m_extensionStatus.ContainsKey(code))
                {
                        if (m_AoInit.CheckInExtension(code) ==
esriLicenseStatus.esriLicenseCheckedIn)
                        {
                            m_extensionStatus[code] = esriLicenseStatus.
```

```
esriLicenseCheckedIn;
                    }
                }
            }
    }

    ///<summary>
    /// Get/Set the ordering of product code checking. If true, check
from lowest to
    ///highest license. True by default.
    ///</summary>
    public bool InitializeLowerProductFirst
    {
            get
            {
                return m_productCheckOrdering;
            }
            set
            {
                m_productCheckOrdering = value;
            }
    }

    ///<summary>
    /// Retrieves the product code initialized in the ArcObjects
application
    ///</summary>
    public esriLicenseProductCode InitializedProduct
    {
            get
            {
                try
                {
                    return m_AoInit.InitializedProduct();
                }
                catch
                {
                    return 0;
```

```
                }
            }
    }
#region Helper methods
    private bool Initialize()
    {
            if (m_requestedProducts == null ||m_requestedProducts.
Count == 0)
                return false;

            esriLicenseProductCode currentProduct = new esriLicense
ProductCode();
            bool productInitialized = false;

            //Try to initialize a product
            ILicenseInformation licInfo = (ILicenseInformation)m_
AoInit;

            m_requestedProducts.Sort();
            if (! InitializeLowerProductFirst) //Request license from
highest to lowest
                m_requestedProducts.Reverse();

            foreach (int prodNumber in m_requestedProducts)
            {
                esriLicenseProductCode prod =(esriLicenseProductCode)
Enum.ToObject(typeof(esriLicenseProductCode), prodNumber);
                esriLicenseStatus status = m_AoInit.IsProductCode
Available(prod);
                if (status == esriLicenseStatus.esriLicenseAvailable)
                {
                        status = m_AoInit.Initialize(prod);
                        if (status == esriLicenseStatus.esriLicense
AlreadyInitialized ||
                            status == esriLicenseStatus.esriLicense
CheckedOut)
                        {
                            productInitialized = true;
```

```
                        currentProduct = m_AoInit.Initialized
Product();
                    }
                }

                m_productStatus.Add(prod, status);

                if (productInitialized)
                    break;
            }

            m_hasInitializeProduct = productInitialized;
            m_requestedProducts.Clear();

              // No product is initialized after trying all requested
licenses, quit
            if (! productInitialized)
            {
            return false;
            }

            //Check out extension licenses
            return CheckOutLicenses(currentProduct);
        }

        private bool CheckOutLicenses(esriLicenseProductCode current
Product)
        {
            bool allSuccessful = true;
            //Request extensions
            if (m_requestedExtensions ! = null && currentProduct ! = 0)
            {
                foreach (esriLicenseExtensionCode ext in m_requested
Extensions)
                {
                    esriLicenseStatus licenseStatus =m_AoInit.IsExtension
CodeAvailable(currentProduct, ext);
                    if (licenseStatus = = esriLicenseStatus.esriLicense
```

```
Available)
                //skip unavailable extensions
              {
                  licenseStatus = m_AoInit.CheckOutExtension(ext);
              }
              allSuccessful = (allSuccessful && licenseStatus ==
esriLicenseStatus.esriLicenseCheckedOut);
              if (m_extensionStatus.ContainsKey(ext))
                  m_extensionStatus[ext] = licenseStatus;
              else
                  m_extensionStatus.Add(ext, licenseStatus);
          }

          m_requestedExtensions.Clear();
      }

      return allSuccessful;
  }

private string ReportInformation (ILicenseInformation licInfo,
esriLicenseProductCode code,
esriLicenseStatus status)
  {
          string prodName = string.Empty;
          try
          {
              prodName = licInfo.GetLicenseProductName(code);
          }
          catch
          {
              prodName = code.ToString();
          }

          string statusInfo = string.Empty;

          switch (status)
          {
          case esriLicenseStatus.esriLicenseAlreadyInitialized:
```

```
            case esriLicenseStatus.esriLicenseCheckedOut:
                 statusInfo = string.Format(MessageProductAvailable,
prodName);
                break;
            default:
                statusInfo = string.Format(MessageProductNotLicensed,
prodName);
                break;
            }

            return statusInfo;
    }
private string ReportInformation (ILicenseInformation licInfo,
esriLicenseExtensionCode code,
esriLicenseStatus status)
    {
            string extensionName = string.Empty;
            try
            {
                extensionName=licInfo.GetLicenseExtensionName(code);
            }
            catch
            {
                extensionName = code.ToString();
            }

            string statusInfo = string.Empty;

            switch (status)
            {
            case esriLicenseStatus.esriLicenseAlreadyInitialized:
            case esriLicenseStatus.esriLicenseCheckedOut:
                statusInfo = string.Format(MessageExtensionAvailable,
extensionName);
                break;
            case esriLicenseStatus.esriLicenseCheckedIn:
                break;
            case esriLicenseStatus.esriLicenseUnavailable:
```

```
                statusInfo = string.Format(MessageExtension
Unavailable, extensionName);
                break;
            case esriLicenseStatus.esriLicenseFailure:
                statusInfo = string.Format(MessageExtensionFailed,
extensionName);
                break;
            default:
                statusInfo = string.Format(MessageExtensionNotLicensed,
extensionName);
                break;
            }

            return statusInfo;
    }
    #endregion
}
```

# 附录 2　Dbf 读写类源代码

```
class DbfOper
{
    string _DirectoryName = "";
    public DbfOper(string strDirectory)
    {
        _DirectoryName = strDirectory;
    }

    public DataTable ReadDbf( string filename)
    {
        string sConn =
            "Provider=Microsoft.Jct.OLEDB.4.0; " +
            "Data Source=" + _DirectoryName + "; " +
            "Extended Properties=dBASE IV;";

        OleDbConnection conn = new OleDbConnection(sConn);
        conn.Open();

        string sql = @ "select * from " + filename;
        OleDbDataAdapter da = new OleDbDataAdapter(sql, conn);
        DataTable dt = new DataTable();
        da.Fill(dt);

        return dt;
    }

    public void WriteDbf(DataTable dt)
    {
        //Console.WriteLine("Writing to: " + dt.TableName + ".dbf ...");
        //如果存在同名文件则先删除
```

```
string fullname = _DirectoryName + "\\" + dt.TableName + ".dbf";
if (File.Exists(fullname))
{
    //Console.WriteLine("Delete file: " + fullname);
    File.Delete(fullname);
}

//建立数据库连接(存储目录名看作数据库名)
string sConn =
    "Provider=Microsoft.Jet.OLEDB.4.0; " +
    "Data Source=" + _DirectoryName + "; " +
    "Extended Properties=dBASE IV;";
OleDbConnection conn = new OleDbConnection(sConn);
conn.Open();
try
{
    OleDbCommand cmd = null;
    //建立新表
    string dbfHeader = DbfHeaderClause( dt );
    //Console.WriteLine(" \nCreating Table ...");
    //Console.WriteLine(dbfHeader);
    cmd = new OleDbCommand(dbfHeader, conn);
    cmd.ExecuteNonQuery();

    //插入各行
    foreach (DataRow dr in dt.Rows)
    {
        string sbInsert = DbfInsertClause(dt, dr);
        //Console.WriteLine(" \nInserting lines ...");
        //Console.WriteLine(sbInsert);
        cmd = new OleDbCommand(sbInsert, conn);
        cmd.ExecuteNonQuery();
    }
}
catch (Exception ex)
{
    Console.WriteLine(ex.Message);
}
```

```
    conn.Close();
}
///<summary>
///建立 Dbf 表头语句
///</summary>
///<param name="dt"></param>
///<returns></returns>
private string DbfHeaderClause(DataTable dt)
{
    StringBuilder sbCreate = new StringBuilder();
    sbCreate.Append("CREATE TABLE " + dt.TableName + ".dbf (");
    for (int i = 0; i<dt.Columns.Count; i++)
    {
        sbCreate.Append(dt.Columns[i].ColumnName);
        sbCreate.Append(" char(25)");
        if (i < dt.Columns.Count - 1)
        {
            sbCreate.Append(", ");
        }
        else
        {
            sbCreate.Append(')');
        }
    }

    return sbCreate.ToString();
}
///<summary>
///插入行语句
///</summary>
///<param name="dt"></param>
///<param name="dr"></param>
///<returns></returns>
private string DbfInsertClause(DataTable dt, DataRow dr)
{
    StringBuilder sbInsert = new StringBuilder();
    sbInsert.Append("INSERT INTO " + dt.TableName + ".dbf (");
    for (int i = 0; i<dt.Columns.Count; i++)
```

```
        {
            sbInsert.Append(dt.Columns[i].ColumnName);
            if (i != dt.Columns.Count - 1)
            {
                sbInsert.Append(", ");
            }
        }
        sbInsert.Append(") VALUES (");
        for (int i = 0; i<dt.Columns.Count; i++)
        {
            sbInsert.Append("'" + dr[i].ToString() + "'");
            if (i != dt.Columns.Count - 1)
            {
                sbInsert.Append(", ");
            }.
        }
        sbInsert.Append(')');
        return sbInsert.ToString();
    }
}
```

# 附录 3　ArcEngine 桌面应用程序框架设计(无模板)

## 1. 新建项目

①单击【文件】→【新建】→【项目】。

②在模板目录上点击 Visual C#，选择 Windows 窗体应用程序，项目名称命名为"WindowsFormsApp"，点击【确定】，生成空白的 Windows 应用程序框架，如附图 3-1，附图 3-2 所示。

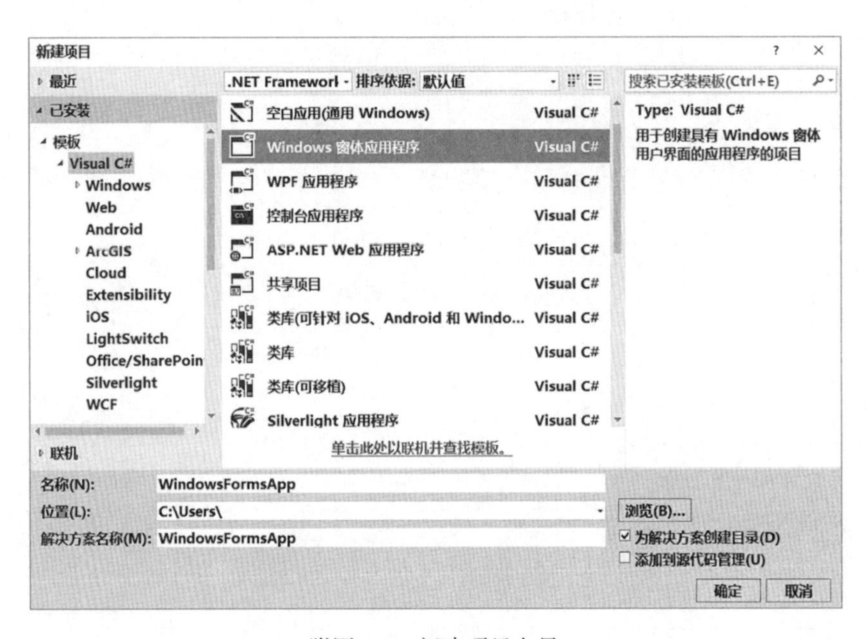


附图 3-1　新建项目向导

## 2. 添加引用

### (1)引用 ArcGIS Engine 组件

右键单击【引用】→【添加引用】，在引用管理器对话框(附图 3-3)中选择：【程序集】→【扩展】，在选项列表中选中以下组件，单击【确定】添加到项目：

①ESRI. ArcGIS. DataSourcesFile；

②ESRI. ArcGIS. Carto；

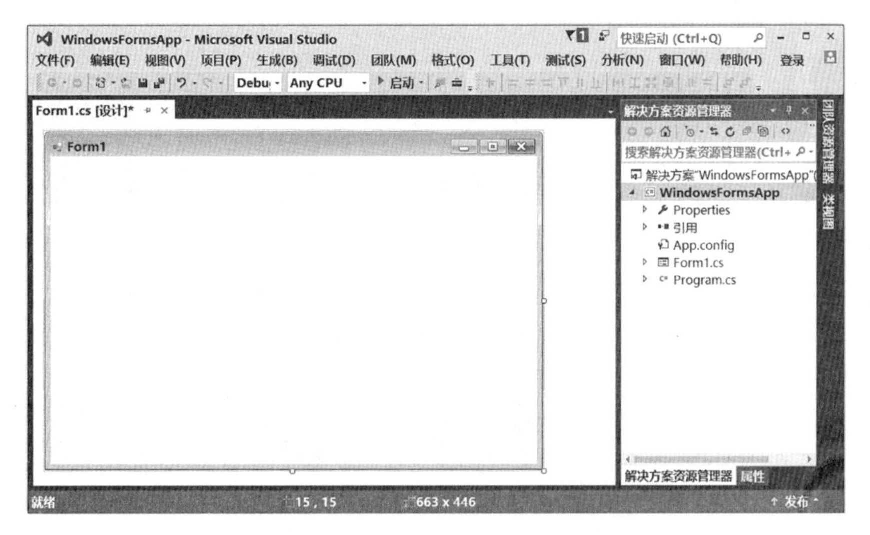


附图 3-2　Windows 应用程序设计界面

③ESRI. ArcGIS. Geodatabase；

④ESRI. ArcGIS. Geometry。

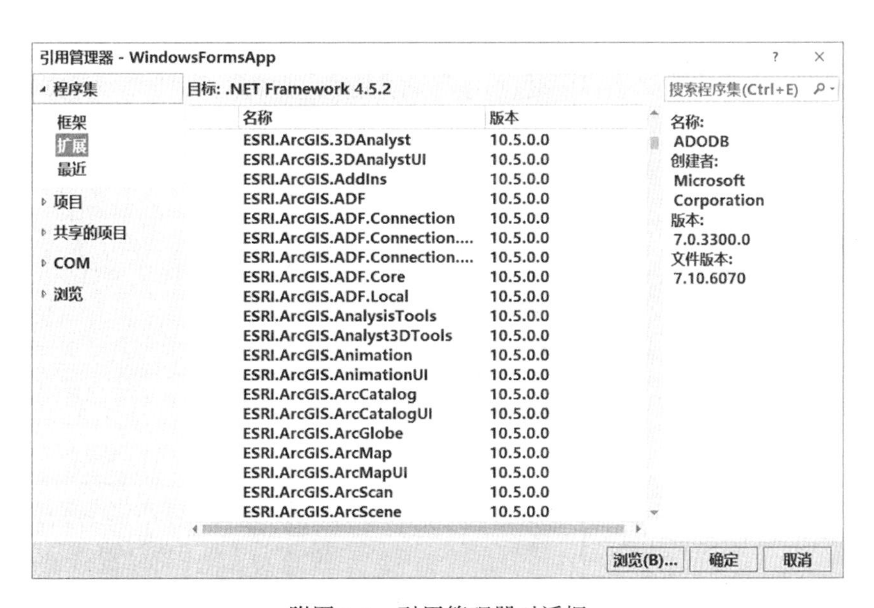


附图 3-3　引用管理器对话框

**(2)该程序所需所有的 using 指定**

①using ESRI.ArcGIS.Carto;

②using ESRI.ArcGIS.DataSourcesFile;

③using ESRI.ArcGIS.Display;

④using ESRI.ArcGIS.Geodatabase;

⑤using ESRI.ArcGIS.Geometry;
⑥using ESRI.ArcGIS.Controls;
⑦using ESRI.ArcGIS.SystemUI.

### 3. 添加 ArcGIS Windows Forms 控件

#### (1)打开工具栏

点击工具箱中的 ArcGIS Windows Forms 工具栏，将看到 MapControl、TOCControl、ToolbarControl 和 LicenseControl 工具，如附图 3-4 所示。

附图 3-4　工具器

#### (2)添加控件并设置停靠属性

将 ToolbarControl 控件拖入 Form1 窗体中，设置 Dock 属性为"Top"(从 Dock 下拉式列表中选择最上方的长条，如附图 3-5 所示。

将 TOCControl 控件拖入 Form1 窗体中，设置 Dock 属性为"Left"(最左边长条)，将工具栏中的 Splitter 控件拖入 Form1 窗体中，设置 Dock 属性为"Left"，这个控件将窗体纵向

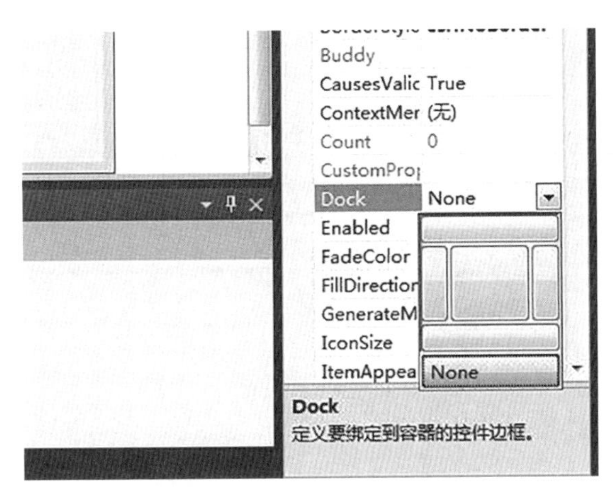


附图 3-5　停靠属性设置

分为两部分。

将 MapControl 控件拖入 Form1 窗体中，设置 Dock 属性为“Fill”(中间的正方形(Fill)，即窗体如附图 3-6 所示。

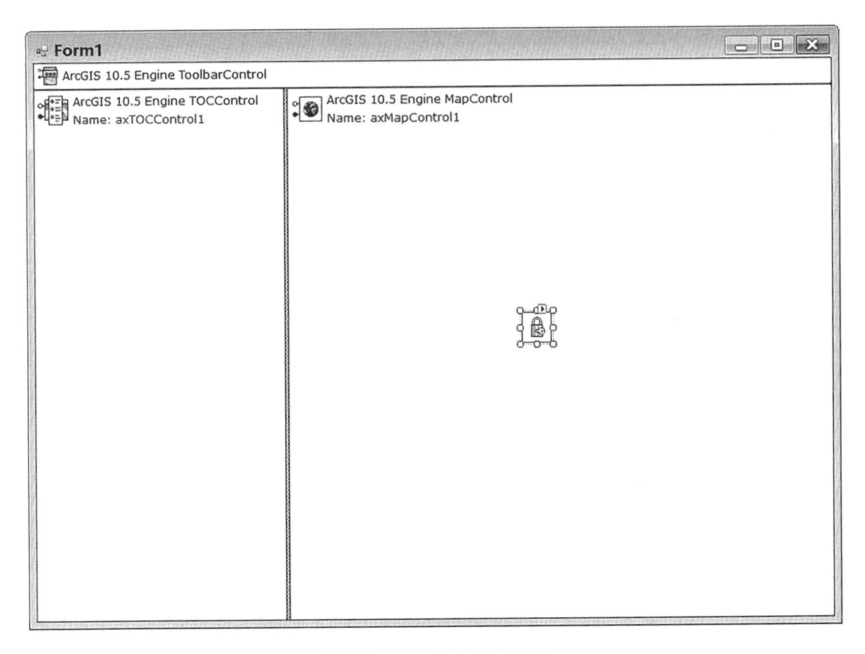


附图 3-6　主窗体布局

**(3)设置绑定**

右键单击 ToolbarControl 控件，在浮动菜单上单击【属性】，弹出属性对话框，如附图 3-7 所示，选择 General 属性，在 Buddy 下拉列表中选择“axMapControl1”，同理，在 TOCControl 属性对话框的 General 属性页，在 Buddy 下拉列表中选择 axMapControl1，点击

【确定】。

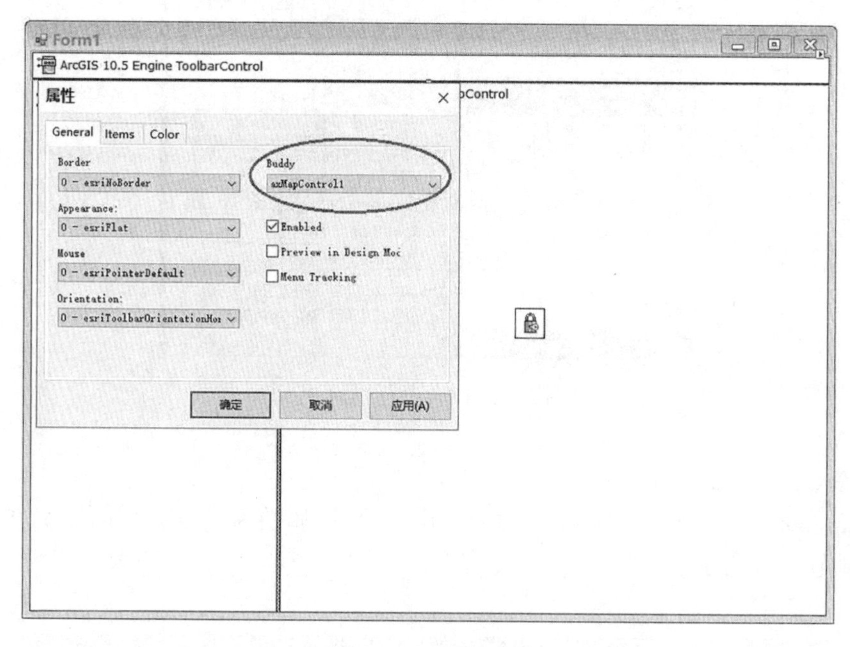


附图 3-7　绑定设置

**(4)添加工具**

在 ToolbarControl 属性对话框中，点击 Items 属性页，单击【Add】，弹出【Controls Commands】对话框，选择 Commands 选项页，左侧单击命令分类，双击右侧命令(或拖动到 Items 容器中)添加工具(如附图 3-8 所示)，点击确定。

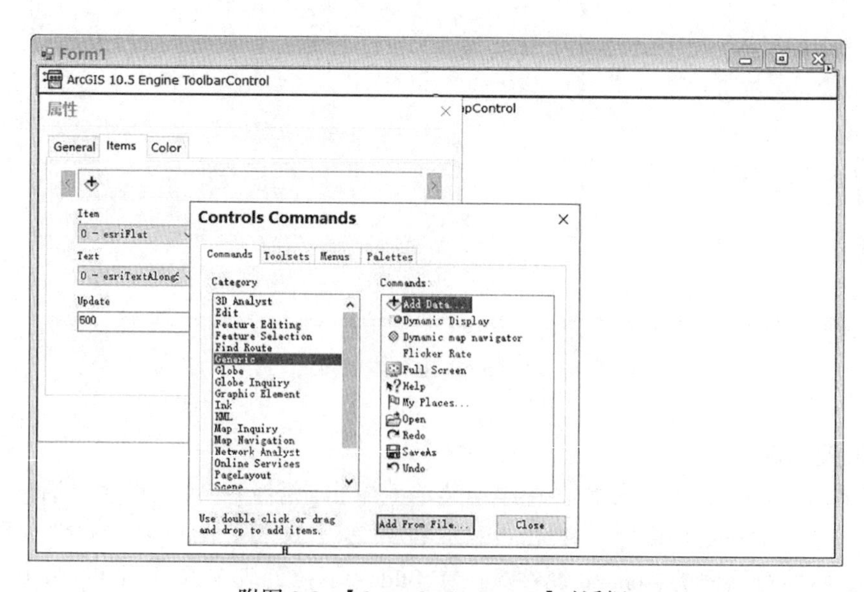


附图 3-8　【Controls Commands】对话框

### 4. 代码添加 MXD

①选择工具箱中菜单和工具栏，将其中 MenuStrip 拖入窗体上方，添加菜单【File】，之后再添加菜单项【Open MXD】，如附图 3-9 所示。

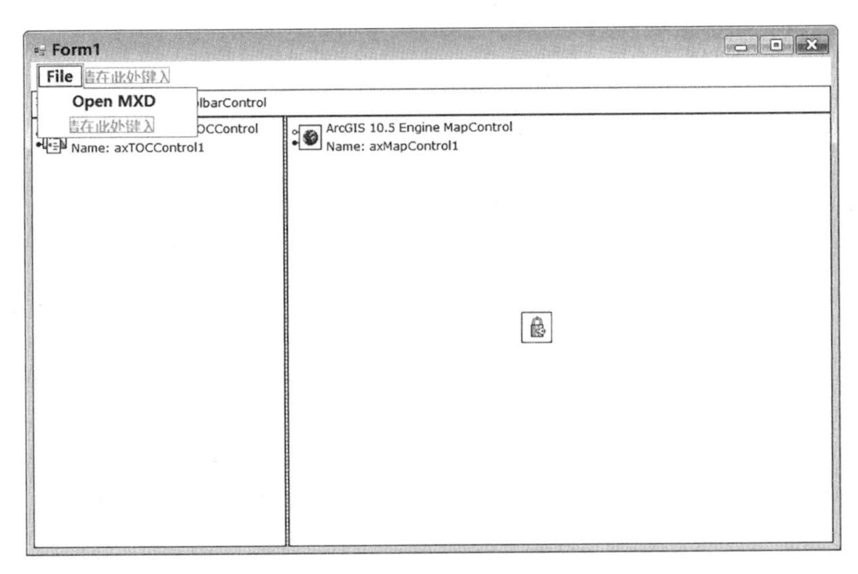


附图 3-9　添加菜单

②右键单击【Open MXD】或双击，向其中输入如下代码：

```
private void openMXDToolStripMenuItem1 _ Click ( object sender,
EventArgs e)
{
    OpenFileDialog OpenMXD = new OpenFileDialog();
    OpenMXD.Title = "打开地图";
    OpenMXD.InitialDirectory = "C:";
    OpenMXD.Filter = "Map Documents ( * .mxd) |* .mxd";
    if (OpenMXD.ShowDialog() = = DialogResult.OK)
    {
        string MxdPath = OpenMXD.FileName;
        axMapControl1.LoadMxFile(MxdPath);
    }
}
```

### 5. 版本绑定

打开 Program. cs 文件，在 Main 函数中添加版本绑定代码(这是 ArcGIS Engine 10. 0 后引入的版本控制模式)，代码如下：

```
static void Main()
{
    if (! RuntimeManager.Bind(ProductCode.Engine))
    {
        if (! RuntimeManager.Bind(ProductCode.Desktop))
        {
            MessageBox.Show("Unable to bind to ArcGIS runtime.
Application will be shut down.");
            return;
        }
    }

    Application.EnableVisualStyles();
    Application.SetCompatibleTextRenderingDefault(false);
    Application.Run(new Form1());
}
```

注意：需要添加 Esri. ArcGIS. Version 引用。

### 6. 编译测试

单击 F5 键开始运行，点击【Open MXD】菜单，在文件浏览器中选择一个 ArcMap 文档（*.mxd），点击【确定】，即可看到运行效果，如附图 3-10 所示。

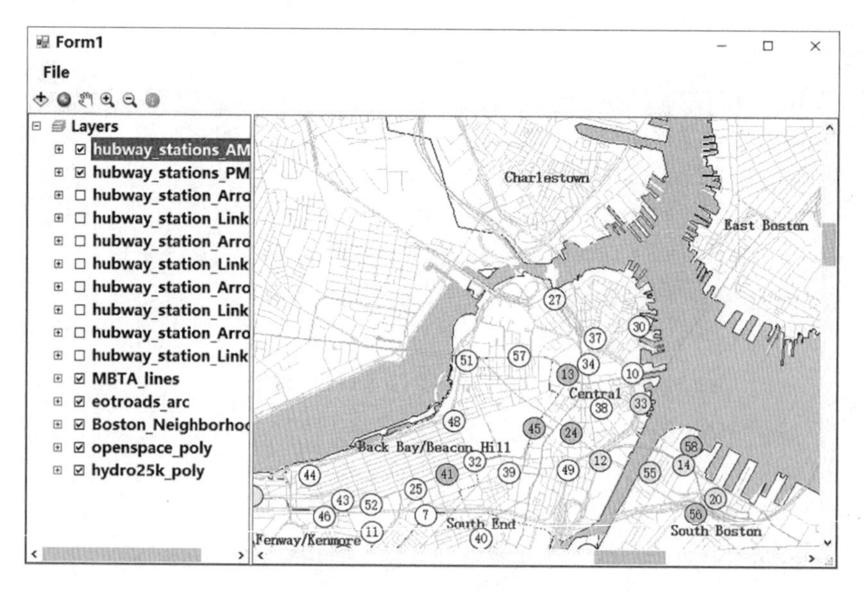


附图 3-10　效果图

### 7. 添加状态栏

此部分由学生自己完成。

### 8. 添加浮动菜单

此部分由学生自己完成。

# 参 考 文 献

1. 汤国安. ARCGIS 地理信息系统空间分析实验教程[M]. 北京：科学出版社，2006.

2. ESRI 中国(北京)培训中心. ArcGIS 轻松入门教程——ArcGIS Engine，2008.

3. 杨克诚. GIS 软件实验指导书[M]. 昆明：云南大学出版社，2009.

4. ESRI 中国(北京)有限公司. ArcGIS 10 产品白皮书，2010.

5. 赵军，刘勇. 地理信息系统 ArcGIS 实习教程[M]. 北京：气象出版社，2011.

6. 张丰，杜振洪，刘仁义. GIS 程序设计教程——基于 ArcGIS Engine 的 C#开发实例[M]. 杭州：浙江大学出版社，2012.

7. 荆平. 基于 C#的地理信息系统设计与开发[M]. 北京：清华大学出版社，2013.

8. 丘洪钢，张青莲，熊友谊. ArcGIS Engine 地理信息系统开发——从入门到精通[M]. 武汉：人民邮电出版社，2013.

9. 刘亚静，姚纪明，陈光. 地理信息系统应用教程——SupperMap iDesktop 7C[M]. 武汉：武汉大学出版社，2014.

10. ESRI. ArcGIS Engine Help For. NET(VS 2015)，2015.

11. 芮小平，于雪涛. 基于 C#语言的 ArcGIS Engine 开发基础与技巧[M]. 北京：电子工业出版社，2015.

12. 牟乃夏，王海银，李丹，等. ArcGIS Engine 地理信息系统开发教程——基于 C#.NET[M]. 北京：测绘出版社，2015.

13. 李进强，基于 ArcGIS Engine 地理信息系统开发技术与实践[M]. 武汉：武汉大学出版社，2017.